building the perfect
gazebo

building the perfect
gazebo

a complete step-by-step guide

DAVID SCHIFF

LARK BOOKS

A Division of Sterling Publishing Co., Inc.
New York

DEDICATION

To my mother, Doris Schiff, who ensures that any book I write is a coffee-table book.

Library of Congress Cataloging-in-Publication Data

Schiff, David, 1955–.
    Building the perfect gazebo : a complete step-by-step
guide / David Schiff.
        p. cm.
Includes index.
    ISBN 1-57990-342-8
    1. Gazebos—Design and construction—Amateurs'
manuals. I. Title.
 TT4963.S35 2004
 690.89— dc21   200310179

10 9 8 7 6 5 4 3 2 1

editor:
JANE LAFERLA

art director:
CHRIS BRYANT

photographer:
INGELA NOREN

cover designer:
BARBARA ZARETSKY

cover photo:
STEPHEN CARVER

illustrator:
FRANK ROHRBACH

gazebo builder:
DANIEL GRANT

technical reader:
DOUGLAS GOODALE

assistant editors:
VERONIKA ALICE GUNTER
ANNE HOLLYFIELD
NATHALIE MORNU

production assistance:
SHANNON YOKELEY
LANCE WILLE

editorial assistance:
DELORES GOSNELL

Published by Lark Books, a division of
Sterling Publishing Co., Inc.
387 Park Avenue South, New York, N.Y. 10016

© 2004, David Schiff

Distributed in Canada by Sterling Publishing,
c/o Canadian Manda Group, 165 Dufferin Street
Toronto, Ontario, Canada M6K 3H6

Distributed in the U.K. by Guild of Master Craftsman Publications Ltd.,
Castle Place, 166 High Street, Lewes, East Sussex, England BN7 1XU
Tel: (+ 44) 1273 477374, Fax: (+ 44) 1273 478606
Email: pubs@thegmcgroup.com, Web: www.gmcpublications.com

Distributed in Australia by Capricorn Link (Australia) Pty Ltd.,
P. O. Box 704, Windsor, NSW 2756 Australia

The written instructions, photographs, designs, patterns, and projects in this volume are intended for the personal use of the reader and may be reproduced for that purpose only. Any other use, especially commercial use, is forbidden under law without written permission of the copyright holder.

Every effort has been made to ensure that all the information in this book is accurate. However, due to differing conditions, tools, and individual skills, the publisher cannot be responsible for any injuries, losses, and other damages that may result from the use of the information in this book.

If you have questions or comments about this book, please contact:
Lark Books, 67 Broadway, Asheville, NC 28801, (828) 253-0467

Manufactured in China

1-57990-342-8

For information about custom editions, special sales, premium and corporate purchases, please contact Sterling Special Sales Department at 800-805-5489 or specialsales@sterlingpub.com.

contents

introduction

I wrote this book with three types of people in mind:

■ Homeowners who may not plan to build their own gazebo, but want to be involved in the design process. If this describes you, I hope you will share this book with your builder and use it as a jumping off point and design tool.

■ Experienced builders who don't need me to tell them how to construct a wooden building, but should benefit from the information specific to gazebos. For example, even if you can design a hip roof in your sleep, you may not have thought lately about how gazebo hips differ from house hips.

■ Creative, ambitious do-it-yourselfers whose favorite part of carpentry is framing. To me, building a gazebo is "sport framing." You're not remodeling a room, creating

indoor dust and debris, and inconveniencing the whole family. You're out there in the sunshine, doing what you enjoy, working at your leisure, creating something beautiful in a nice, comfortable scale. Building a gazebo is the perfect opportunity to stretch your carpentry chops, and I designed this book to help you do that.

No matter which type of gazebo-lover you are, I hope you have as much fun creating your own gazebo as I had creating this book.

David Schiff

How
to Use
This Book

You will not find a collection of plans for pre-designed gazebos in this book. That's because I am not really interested in helping you build someone else's gazebo. I would much rather help you design and build *your* gazebo. I am excited by the idea of helping you create something that expresses your style and taste and perfectly complements your unique outdoor environment.

For example, rather than giving you the floor framing plan for one specific eight-sided gazebo, I discuss the pros and cons of a concentric floor frame versus a doubled-beam floor frame, so you can decide which one you want to build. Rather than tell you what the dimensions are for some specific gazebo rafter, I give you the information you need to design any rafter you want. Along the way, I've filled the book with options and examples for everything from foundations to railings to cupolas, always giving you the pros and cons of each approach and discussing how it interacts with other options. The first three chapters will help you plan your gazebo, while chapters four through nine will help you build it.

Now, we did build one gazebo specifically for this book—it's that wonderful gazebo with a lookout tower and sunburst railing that you see on the cover.

One reason we built it was to have an opportunity to photograph many of the procedures you'll find throughout the book. But the main reason was to provide me with a vehicle for writing the last chapter: One Gazebo Start to Finish.

After presenting all the design and building choices you'll make in creating your gazebo, I wanted to pull it together into a step-by-step procedure for one specific gazebo. If, as I hope, you've developed your own creative vision of a gazebo, the last chapter will work as a template to help you plan your order of work.

Of course, if you fall in love with the gazebo on the cover, the last chapter will give you the drawings, plans, and sequences you need to replicate it. If you need more details for those sequences, you'll find them cross-referenced in pertinent sections throughout the book. To help you find them, look for this icon.

The classical lines of this gazebo in Brookline, Massachusetts, make it look like a temple from another time. PHOTO BY DICK DIETRICH.

CHAPTER ONE

A Brief History of Gazebos in America

People of means have always enhanced their gardens with decorative structures. In ancient Egypt about 3,500 years ago, King Amenhotep III liked his garden enough to want to take it with him to the afterlife. On his tomb is a mural of his garden that includes four small-roofed structures to shade him from the sun as he gazes at his pools.

The Early American Summerhouse

Through most of American history, gazebos continued as a garden feature reserved for the well-to-do. George Washington had a small eight-sided garden structure at Mount Vernon. Thomas Jefferson wrote about gazebos—then usually called summerhouses or pavilions. He planned to build at least three in different styles at his home, Monticello, but only got around to building one. (Jefferson's plans were always bigger than his purse.) He reportedly designed one for James Madison's house at Montpelier, Virginia.

Early American summerhouses weren't the fanciful wooden structures that the word gazebo is likely to bring to the modern mind. Since the wealthy built them, early garden structures tended to be substantial and elaborate, most often constructed of brick or stone. In fact the word gazebo didn't exist until the mid-18th century; the earliest reference the editors of the *Oxford English Dictionary* found was in 1752. But no one knows for sure where the word came from. The Oxford editors suggest it is a "humorous formation" combining the word gaze with the Latin word *videbo* which means "I shall see."

In early usage at least, the difference between a gazebo and summerhouse derived from the structure's intended use: Gazebos should have a commanding view of the countryside or garden, whereas a summerhouse can be any small building in the garden. Belvedere, from the Italian words for beautiful view, was a term often used to de-

A Chinese pagoda, depicted here as a motif in a tapestry, could serve as a plan for a fanciful garden structure.
PHOTO COURTESY, WINTERTHUR MUSEUM.

Thomas Jefferson's brick summerhouse at his home, Monticello, is an ancestor of our modern concept of the gazebo.   PHOTO COURTESY, MONTICELLO/THOMAS JEFFERSON FOUNDATION, INC.

This gazebo complements the James P. White House in Belfast, Maine. Built in 1840, the house is an excellent example of Greek Revival architecture. PHOTO BY BRIAN VANDEN BRINK.

scribe a summerhouse with a view before the word gazebo was coined.

Early Americans of means were usually well educated in Latin and the classics. Familiar with Roman architecture, they imitated it extensively in their homes and outbuildings. It was not uncommon for summerhouses or pavilions to look like small Roman temples complete with classical stone columns.

Strongly influenced by British fashion and culture, wealthy Early Americans adopted the British craze for objects in the Oriental style, which came to be known as *chinoiserie*. Fanciful gazebo-like structures were often depicted in wallpapers and fabrics inside the home and sometimes got built as actual structures in the garden. Concave roofs and roofs with two levels, suggestive of Chinese pagodas, continue to be popular gazebo details to this day, as do Chinese-style brackets and railings.

The very wealthiest Early Americans indulged themselves in the so-called *natural style* that was becoming popular with the British titled class. This style eschewed formal geometric gardens in favor of informal groupings of trees and shrubs, long magnificent views and, if possible, water features such as ponds or even lakes. Long meandering walks were an essential part of any natural landscape, and along the walks summerhouses provided a shady place to rest.

Besides requiring large amounts of money for development, the natural style required huge tracts of land. Fewer people in America than in England possessed these combined means, making the natural style less prevalent here. However, the natural style strongly

influenced the designs of Washington's Mount Vernon and Jefferson's Monticello.

During the early part of the 19th century, Greek architecture became the dominant influence in American architecture; it seemed only natural that this fledgling democracy would identify with that ancient one. The Greek building forms of columns, cornices, and dentil moldings were simpler than in the Roman style, lending themselves readily to interpretation in wood. In the United States today, you can see the Greek Revival style in grand government buildings as well as in humble 19th-century farmhouses. Naturally, if your house was Greek Revival in style, so was your icehouse, privy, and, of course, your gazebo.

Built in 1835 at the Rosedown Plantation in St. Francisville, Louisiana, this gazebo is a fine example of Greek Revival style executed in wood lattice. PHOTO BY DICK DIETRICH.

The Rustic and the Picturesque

In 1841, a young nurseryman named Andrew Jackson Downing published a book titled *A Theory and Treatise on the Theory and Practice of Landscape Gardening, Adapted to North America; with a View to the Improvement of Country Residences.* This book was so successful that many consider Downing to be the most influential writer on American landscape design to this day. He went on to write several other books, and in 1846 became the founding editor of *The Horticulturist* magazine.

Having studied the English natural style of landscaping, Downing professed the idea that we should go beyond imitating nature. Instead, he believed that we should interpret nature. "In landscape gardening," he wrote in the *Treatise,* "we should aim to separate the accidental and extraneous in nature, and to preserve only the spirit, or essence." This pivotal thought was one reason why Downing's writings were so popular. He showed that by selecting the best of what nature had to offer, you didn't necessarily need huge tracts of land to have a picturesque natural landscape.

In the *Treatise* and in later writings, Downing differentiated between what he called the *beautiful* and the *picturesque* in the landscape. He characterized the beautiful as smooth and regular forms such as "some gently undulating plain, covered with emerald turf." The picturesque, which he characterized as "striking irregular, spirited forms," was the style Downing

preferred. An example of the picturesque would be a deep craggy ravine with twisted lark trees hanging from its sides, casting dark dramatic shadows. His love of irregular, rough forms led Downing to advocate rustic furnishings and structures for the garden. He liked the idea that a resourceful person could build fencing, seating, arbors, and even bridges and gazebos using rough-hewn limbs of whatever wood was on hand.

The rustic style continued to be popular even when milled lumber became readily available. Ironically, by the late 19th century, rustic style furniture was being mass-produced in kits that could be delivered and easily assembled on site. Even more ironically, the dawn of the industrial age meant you could buy rustic-style benches, fences, and gazebos made from iron.

During the latter part of the 19th century, ease of travel made possible by the railroads led wealthy denizens of New York City to establish vacation "camps" in the Catskill and Adirondack Mountains. Inspired by the rustic structures that Fredrick Law Olmsted designed for Manhattan's Central Park, these camps were built in the rustic style. Soon rustic Adirondack resorts sprung up for people who couldn't afford to build their own camps. Rustic gazebos were always part of these resorts and camps. One resort, the Mohonk Mountain House in New Paltz, New York, in the Catskill Mountains, is still open today and features more than 100 summerhouses.

The Victorian Era

The taste for things Victorian began to emerge around the middle of the 19th century, but it wasn't until the 1870s that the Victorian rage really took hold in America. A number of converging factors fueled this eclectic explosion after the Civil War: Wood was abundant and cheap; new machines made it possible to mass-produce ornate milled moldings and brackets; and the railroads shipped building materials, which cut the time and cost of transport. Most importantly, an emerging middle class meant the average Joe had a few extra dol-

Built on a rocky mountainside at the Mohonk Mountain House, New Paltz, New York, this gazebo is accessible only by a rustic bridge. It is a perfect example of Andrew Jackson Downing's ideal landscape. PHOTO COURTESY, MOHONK MOUNTAIN HOUSE.

lars in his pocket and a few leisure hours in his week.

As transportation and mail service improved, magazines became an increasingly popular source of information. Then, as now, many of these magazines were about gardening, which was becoming what it is today—one of America's favorite leisure-time activities. Like today, the gardening magazines featured articles about outbuildings, including gazebos. With this information, anyone with a little yard around his home could build a gazebo.

Of course, the Victorian fellow didn't pay someone to design his landscape and outbuildings. More likely, he found inspiration from a picture or from his neighbor's garden. Maybe he built the gazebo himself or hired a local carpenter to adapt a plan he saw in a book or magazine. (Sound familiar?) In the eclectic Victorian spirit, he wasn't overly concerned that the gazebo faithfully replicate the architectural style or period of his house.

The post-Civil War building boom also brought the development of the *balloon frame*. This system of house framing used long, thin pieces of lumber that could be assembled with nails much more quickly and with less skill than tra-

ditional timber framing, with its labor-intensive mortise-and-tenon joinery. The balloon frame, along with readily available nails and lightweight lumber, made it easier than ever to create a small garden structure. And so the gazebo, as we think of it today, became common in American gardens: a whimsical wooden structure replete with fancy Carpenter Gothic railings and brackets.

Gazebos for the 21st Century

The popularity of gazebos has waxed and waned over the past 100 years. By the turn of the 20th century, they were falling out of fashion. People were tiring of the intense ornamentation of the Victorian style, and perhaps thought of gazebos as part of the bric-a-brac. Furthermore, it had become fashionable to enjoy the outdoors from the vantage of a wide front porch. Ironically, gazebos made a brief comeback during the 1930s when the noise and bustle of automobiles drove people off those same front porches. In the suburban building boom after World War II, the backyard patio became the venue of choice for outdoor respite. By the 1970s, the outdoor deck was beginning its reign of popularity.

The mid-1970s saw a United States that was weary of the turmoil of the Vietnam war era. Concurrent with the nation's bicentennial, people were nostalgic for what they perceived as simpler times. Historic preservation gained momentum, encompassing both Colonial and Victorian structures. As the cost of new housing soared through the 1980s, many young people, myself included, turned their attention to the stock of affordable rundown, Victorian-era fixer-uppers in small towns throughout the country.

We looked past the leaky roofs and drafty windows to the high ceilings, spacious rooms, and won-

In a beautiful marriage of old forms and new, this nostalgic gazebo meets a deck near Madison, Wisconsin.
PHOTO BY JESSIE WALKER, COURTESY OF THE CALIFORNIA REDWOOD ASSOCIATION.

Ron Forest built this gazebo with color and siding to match his new house in Scarborough, Maine. The gazebo's turret roof matches several turrets on the house. Its arched openings with key blocks are a subtle nod to classical Greek architecture.
PHOTO BY BRIAN VANDEN BRINK.

derful moldings of these white elephants. With a couple of years and a little elbow grease, we could have a home that would far surpass anything you could buy new. (I guess it was a good thing we were initially so naïve about the work, time, and money involved.) When the work was finally done, around 1990 or so, someone said, "Wouldn't it be just perfect to build a gazebo back there?"

Today, the interest in gazebos continues unabated as a feature in new homes as well as old. One reason may be that many new houses are built in a style that might be called Victorian Revival. One of my favorite adaptations is a modern Victorian-style house with a deck that has a gazebo built on it. But you need only look around to see that a renewed interest in Victorian style is only a small part of the story.

A simple, quiet little garden retreat, a place where we can enjoy a breeze and perhaps a view as we relax and gather our thoughts, offers an appealing respite to our high-tech, high-speed, 21st-century lives. The great thing is, like the Victorians themselves, 21st-century homeowners are taking an eclectic approach. They realize that this adaptable form we call a gazebo doesn't need to mimic the style of their house—as long as it complements their home and their garden.

Partially hidden by vines, this gazebo is a peaceful private garden retreat.
PHOTO BY CHARLES MANN, GARDEN DESIGN BY DULCY MAHAR.

CHAPTER TWO

Choosing a Site for Your Gazebo

What first gave you the idea you might like to build a gazebo? Perhaps you walked out to a corner of your property, were struck by a magnificent view, and said to yourself, "Wow, this would be the perfect spot for a place to sit!" Well then, you have already located your gazebo. Or maybe you don't have a particular spot in mind, and just think about you would like a lovely garden retreat someplace in the yard.

In either case, it's important to step back and think about how your gazebo will affect the way your property looks and the way it functions. A gazebo that has been plunked down in a yard with no real reason to be there is a sad sight, of use to no one beyond the spiders spinning webs between its posts.

Start by thinking about how you will use the gazebo. Is it a place where you will entertain? Will you dine there? A dining table and chairs require a

What could be more pleasant than a meal in a garden gazebo? Just be sure to plan enough room for easy access around the chairs. **PHOTO BY TERRY WILD**.

bigger gazebo than you may think. You need about 46 inches of space between the edges of the table and the railing so people can walk behind the chairs and pull them out to sit. That means at least a 12-foot- wide gazebo for a typical table that seats four. If possible, take the table and chairs out to the spot where you plan to put the gazebo—this will give you an idea of how big the gazebo should be and how the table will work in the space. And, of course, a gazebo that will be used for dining shouldn't be too far from the kitchen.

The view *of* your gazebo can be as important as the view *from* your gazebo. Do you want your gazebo to be an important focal point that draws the eye when you look out the window or enter the yard? Perhaps it is at the juncture of two or more garden paths. The multi-sided nature of gazebos makes them a natural for placement in the center of things, perhaps with flower beds radiating from the base. You may want to paint your gazebo white or a combination of bright colors to make it stand out. You might even add dramatic uplighting around the base so you can enjoy an evening view of your gazebo.

Will your gazebo be a small, hidden treasure—a place for quiet conversation or solitary medita-tion tucked away in a nook of your yard? Rustic-style gazebos that

A coat of white paint highlights this gazebo's light and airy look. Paint a gazebo white only if you want it to be a focal point in your landscape.
PHOTO BY CHARLES MANN.

blend into the landscape work well this way. Don't make the mistake of thinking that an unobtrusive gazebo won't impact the rest of the yard. Folks will literally beat a path to this new feature. You might even want to pave that path.

Keep in mind that your gazebo may also block a view, which in some cases may be desirable. A strategically placed gazebo, surrounded on two or three sides with tall plantings, is a great way to create a romantic nook while obliterating an undesirable sight. If you want to block an obtrusive view without stifling breezes or cutting too much light, install lattice on several sides of the gazebo.

In the preceding chapter, I gave you a whirlwind tour of the history of gazebos in America. If you have a period home or just want some perspective, you might want to do some more reading about American landscaping. Three books that you may find helpful are *Furnishing the Old-Fashioned Garden* by May Brawley Hill (Harry N. Abrams, Inc., 1998); *The New Traditional Garden* by Michael Weishan (Ballentine Publisher Group, 1999); and *Landscapes and Garden for Historic Buildings* by Rudi J. Favretti and Joy Putman Favretti (AltaMira Press, 1997). And, of course, Andrew Jackson Downing's books are still in print.

This gazebo in Northern California has lattice panels on five of its eight sides to provide shade and privacy without blocking breezes.
PHOTO BY KARL RIEK, COURTESY OF THE CALIFORNIA REDWOOD ASSOCIATION.

Drawing a Site Plan

If you're lucky, you'll have the opportunity to build your gazebo as part of an overall landscape plan for your home. You, or a landscape architect or designer, will draw a site plan that includes all plantings and structures. Even if you are fitting your gazebo into an existing landscape, drawing at least a rough site plan usually is the best way to envision how the gazebo will function with the rest of your yard.

You may have noticed that I said a site plan *usually* is the best way to plan how your gazebo will work in the yard. To be honest, there are some situations in which a site plan is not all that useful.

If the gazebo will be far from the house and other structures and the area around it will not be landscaped, then you probably don't need a site plan. As it happens, the gazebo we built for this book is located on just such a site. We placed it in a meadow of tall grass next to a pond. The plan is to allow natural growth with no additional landscaping. To complement the look, we selected natural-finish cedar for building the gazebo.

However, not drawing a site plan doesn't mean we didn't carefully consider the site. We wanted the gazebo to have some presence, so we built it tall to complement the tall grass and to offset the gazebo's position at the bottom of a hill. In fact, while we were building the gazebo, an immense willow tree behind the house came down in a storm. Suddenly, we had a completely new, unobstructed view of the meadow and gazebo from the house. When we realized the gazebo was going to be a focal point, we decided to make the gazebo even taller than originally planned, and even added a little second-story lookout.

If your gazebo will be located in a natural setting as is the gazebo built for this book, you may find that a site plan is unnecessary. PHOTO BY STEPHEN CARVER.

To make your own site plan, find the survey of your house. If you can't find a copy, you should be able to get one at town hall. The survey will save you time in measuring because it defines the boundaries of your property and locates the house and buildings within those boundaries. It should also locate your driveway, and, if you are lucky, may include other hardscaping features such as walks.

Since a survey doesn't include landscape features such as trees, shrubs, or garden beds, you will need to add them as you create your plan. If for some reason you can't find a copy of your survey, you will have to measure the boundaries of your property and locate the buildings yourself. I'll tell you how to do that a little later.

Let's use my own property as an example of how to use a survey drawing to create a site plan (see figure 1). My wife and I have always thought it might be nice to build a gazebo in a raised area we call the "grassy knoll."

Start by tracing your survey. As you can see from figure 1, I live on a fairly narrow 2.6-acre lot. Since I have an open field to the left of the barn and a vegetable garden and orchard to the right of the shed, I decided to crop those areas out of the site plan and include only the landscaped parts of my property. Be sure to include on your tracing the arrow and line indicating north—it will be important later when you analyze the sun's path over your gazebo.

After tracing the landscaped portion of my yard, I cut it out, centered it from top to bottom along one short edge of a sheet of 8½ x 11-inch paper, and taped it. This ensured that when I set the photocopier for 200 percent, the whole drawing would fit on the paper. Since the scale of the original survey showed 1 inch equal to 60 feet, the scale for my site plan became 1 inch equal to 30 feet. I made a few extra copies in case I made mistakes or wanted to try different ideas.

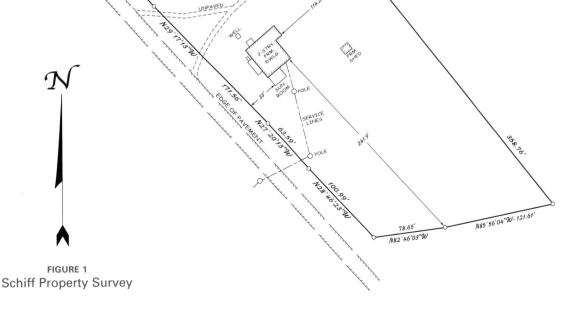

FIGURE 1
Schiff Property Survey

plotting landscape features

The next step is to use triangulation to plot points in your yard that allow you to locate the major features. To do this, you'll need a 100-foot tape measure, a helper, and a compass. A construction calculator that works in feet and inches will save you converting back and forth from decimals. You might want to invest in one if you haven't already; it will come in handy later if you plan to design a roof for your gazebo.

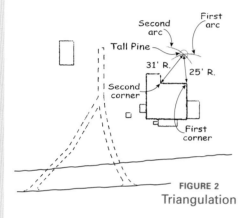

Triangulation

The first thing I plotted in my yard was a tall pine tree that supports one end of my hammock. I measured from one corner of the house to the tree, a distance of 25 feet. To convert 25 feet to the scale of my site plan, I divided 25 inches by 30 to get $1^{3}/_{16}$ inches. I set my compass to this radius, set the pivot point on the same corner of the house on the site plan, and drew an arc. Then I measured from a second corner of the house to the tall pine, a distance of 31 feet. That scales down to a radius of $1^{1}/_{16}$ inch. With the compass set to $1^{1}/_{16}$ inch, I set the pivot point on the second corner of the house and drew another arc, allowing me to see that the tall pine was located where the two arcs intersected as shown in figure 2.

Continue using triangulation in this way to locate key points in your yard—major trees, edges of garden beds, and pathways. Be sure to add any structures that aren't on the survey. If you are designing a landscape from scratch, you may want a more detailed site plan that identifies all the planting species. Remember, you are just trying to get a sense of how your gazebo will fit into an existing picture. You needn't be very accurate in your measurements—within 6 to 12 inches is fine. Don't get carried away plotting irregular shapes such as garden beds; just locate them roughly and estimate their shape, as shown in figure 3.

If you don't have a copy of your survey, pick a couple of points along the perimeter of your property—perhaps the mailbox and a telephone pole. Use these to locate a corner of the house by triangulation. Then measure the perimeter of your house, convert it to scale, and draw it on your site plan.

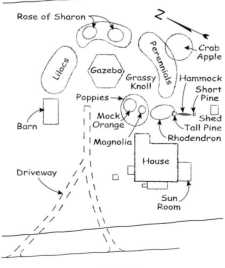

FIGURE 3
A Finished Site Plan

the evolving gazebo

Patricia and Robert Lundquist were novice gardeners with a vision when they moved into their home in Bellingham, Washington, in 1974. Their goal was to create a beautiful retreat on their large property that they dubbed Lundquist Woods.

The Lundquists purchased their small, eight-sided gazebo in 1980, and sited it with its back at the edge of a garden that, over the years, has grown into a shade garden as shown in figure 4. The entrance faced west into a sunlit lawn as an invitation for strollers to sit inside at a small bistro table in the shelter of the slatted roof.

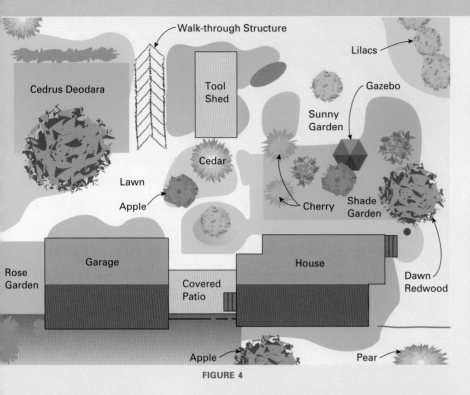

Walk-through Structure

Lilacs

Cedrus Deodara

Tool Shed

Gazebo

Sunny Garden

Cedar

Lawn

Apple

Cherry

Shade Garden

Rose Garden

Garage

House

Covered Patio

Dawn Redwood

Apple

Pear

FIGURE 4

"The gazebo became the true focal point for this retreat, and the garden grew up around it," Patricia Lundquist said. "The west side remains sunny, and the lawn area has become smaller as we created new gardens."

As the gardens evolved, a problem arose. A dwarf box hedge planted on both sides of a path leading into the shade garden's seating areas seemed to dead-end in an unwelcoming fashion at the back of the gazebo. To solve the problem, the Lundquists simply removed the railing and lattice panel from the back of the gazebo directly opposite the sunny garden, creating an entrance from the shade garden.

To showcase the gazebo, the Lundquists altered the curve of the hedged path at both ends. The gazebo now invites visitors from both the sunny and shady sides of the garden and is either a walk-through, or a place to sit for a cool lemonade or iced tea.

The gazebo was first painted white, but a few years later the color was changed to a warm cream to match the trim on the house. In the summer, a lovely wind chime hangs from the top point inside, and baskets of impatiens fill the remaining six "windows" with arresting color.

TOP OF PAGE: With its back removed, the Lundquists' gazebo creates a lovely transition from the sunny garden shown here to the shady garden beyond.
PHOTO BY MARK TURNER.

a bit of the orient in a new jersey suburb

From the front, it's not much different from the other homes in Fair Lawn, a New Jersey suburb of nice houses on small lots within easy commuting distance of New York City. But stepping out Jerry Goldberg's back door is like stepping into another world. A stone walk meanders along a stream. At one end of the stream is the soothing splash of a small waterfall. At the other end is a small pond that disappears under a gazebo whose details evoke Oriental architecture. All around are lush plantings. Sitting back in this gazebo, the pressures of a busy day just seem to slip away. You would never imagine that the neighbor's house is but a few yards away.

Replacing the railings and adding brackets and frie transformed a non-descript, stock gazebo into an Oriental-style retreat. PHOTO BY ALAN AND LINDA DETRICK.

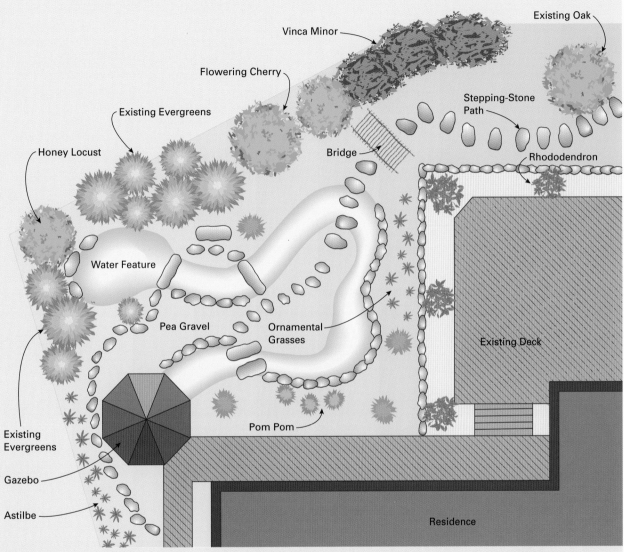

Existing Oak

Vinca Minor

Flowering Cherry

Existing Evergreens

Honey Locust

Stepping-Stone
Path

Bridge

Rhododendron

Water Feature

Pea Gravel

Ornamental
Grasses

Existing Deck

Existing
Evergreens

Gazebo

Astilbe

Pom Pom

Residence

FIGURE 5

This Oriental-style oasis was created by landscape architect David Delardi of Landscape Perceptions, a Wayne, New Jersey, firm. Delardi's site plan (see figure 5) shows the corner of the yard that contains the gazebo, water features, and stone paths. The yard already had a large deck for entertaining, but Delardi decided to add the gazebo to provide a covered area where people could enjoy being in the new space instead of just overlooking it.

"Also, it is a very large water feature," Delardi said. "I wanted something in the same scale to balance the water feature. I wanted to frame the space. I used [the gazebo] as a spatial barrier to block the view of the other property."

As you enter the water garden from the side of the house, the gazebo is the first thing you see. "I also wanted it to act as a gateway into the space," Delardi added. Rather than build the gazebo on site, Delardi bought a prebuilt, 8-foot, eight-sided gazebo.

considering the sun

Creating a place to enjoy beautiful sunsets can be a deciding factor in your gazebo's location. This graceful gazebo is not only a great viewing spot, it also enhances the view. PHOTO BY DICK DIETRICH.

There are several reasons to consider the seasonal and daily path of the sun when siting your gazebo. If, for example, you want your gazebo to be a shady retreat on summer evenings, you may want to orient the gazebo to the east of tall trees so you can enjoy their shade. If your gazebo won't be near trees, you might want to put the entrance facing away from the sun at that time of day and year. This way, the sun will be at your back instead of in your face when you place chairs or benches in the gazebo. Conversely, if the gazebo will have a wonderful view of sunsets, you'll want to orient it toward the sun.

Being aware of the sun's path may be important in the design of your gazebo's roof. Perhaps you are considering building a slat roof instead of a solid roof. You may want to anticipate the pattern of shadows those slats will cast on the gazebo floor.

Speaking of shadows, don't forget that your gazebo will cast one of its own. You don't want to build a gazebo, only to discover your carefully crafted vegetable garden or bed of sun-loving plants is now robbed of summer sunlight. The shadow pattern the gazebo casts in winter is less likely to affect your siting decision, but, still, you might want to anticipate how the shad-

ows will fall on the snow-covered ground on a bright winter day.

To calculate the exact path of the sun and the exact size of the shadows it will cast, you'd need to use a sun chart that gives information for your latitude, since your position on the globe affects the relative height of the sun. But that kind of specific information is really only important if you are designing the orientation and window positions of a house. For a garden structure such as a gazebo, it's enough to know the basic path of the sun, as illustrated in figures 6 and 7. Just in case you picked this book up in a used bookstore in New Zealand, note that these patterns only apply to the Northern Hemisphere.

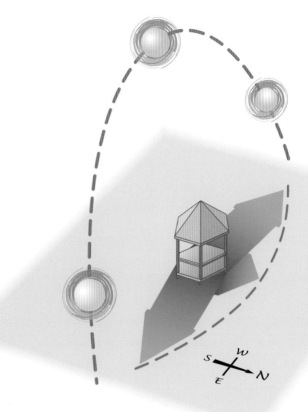

FIGURE 6
Summer Sun Pattern

As we all know, the sun rises in the east and sets in the west. More accurately, on June 22—the longest day of the year, and the summer solstice—the sun rises just slightly northeast and sets just slightly northwest. The result is a long western shadow as the sun rises. The shadow gets shorter until it virtually disappears at noon and then gets progressively longer to the east as the sun moves down toward the west.

After June 22, the sun rises and sets a little farther to the south each day. And so, each day the shadow will get a little bigger on the north side of your gazebo until December 22, the shortest day of the year, and the winter solstice. This means your summer veggies and flowers will get sun all day long, even if you plant them right against the southern side of the gazebo, and they'll still get plenty of sunlight if you plant them to the north. Anything planted to the east or west will have to make do with half a day of direct sunlight.

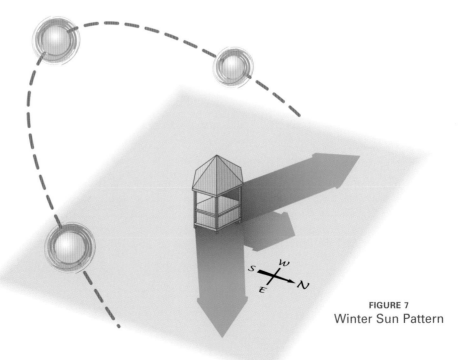

FIGURE 7
Winter Sun Pattern

CHAPTER THREE

Choosing Materials

With a clear idea of what you want your gazebo to look like and how you want it to function, it's time to choose the materials. Since the most important decision is selecting the wood, this topic will be the first and most detailed of this chapter. Next we'll look at wood finishing and painting options, then roofing, and finally the fasteners and metal connectors you'll need to put the whole thing together.

A coat of preservative was all the finish this cedar gazebo needed to blend beautifully in its environment, allowing the bank of heather to take center stage. PHOTO BY MARK TURNER.

Wood for Gazebos

There are several things to think about when choosing wood for your gazebo. For starters, look at what's available in your area. How do those prices compare with other choices? You may get a better deal on local or regional species that don't have to be shipped across the country or the globe. Will you paint your gazebo? Stain it? Leave it natural? You can paint virtually any wood you would use to build a gazebo, but there's no point in paying a premium for a beautiful wood if you only intend to cover it with paint. Also, consider that some woods look and perform better when left to weather naturally.

Most likely, your gazebo will be built from several species of wood.

For example, for the posts and floor framing you might choose pressure-treated, southern yellow pine for strength, economy, and to protect them from wood-eating termites and fungi. The floor might be maho-gany tongue-and-groove porch flooring, chosen for its durability and dimensional stability. Railings and trim boards might be redwood or cedar—woods that weather beautifully. In addition, if you purchase fancy brackets or finials, they may well be made of poplar, a wood that produces crisp details when turned or shaped by machines. Let's take a closer look at the types and grades of wood, so you can choose the best ones for your needs.

Students of the Pennsylvania Institute of Technology in Williamsport, Pennsylvania, built this Victorian-style gazebo on campus using modern construction materials. They achieved the look by using turned balusters, bracket spokes, and a traditional color scheme.
PHOTO COURTESY OF THE PENNSYLVANIA INSTITUTE OF TECHNOLOGY.

wood density

The real key to how well wood stands up to extremes of wet and dry weather is density. Woods that are heavy and strong have more densely packed cells—the tubes that transport moisture-laden sap up the tree and store the tree's food supply. When weather is wet or humid, moisture soaks into the cut ends of the cells (the end grain), and causes them to swell. The denser the wood, the less space there is to accommodate this swelling. As a result, woods that are dense, heavy, and strong will swell and shrink more in response to moisture than lightweight woods that are structurally weaker. Denser woods are also more likely to split, cup, and twist in response to moisture changes.

In general, hardwoods such as oak, cherry, and walnut are denser than softwoods. Hardwoods are rarely used in outdoor construction because they are so dense you can't even drive a nail into them without bending it. And hardwoods are also quite expensive.

Among the softwoods, there is a wide range of density. For example, since it's one of the strongest woods that can be easily nailed, Douglas fir is an excellent choice for load-bearing members such as floor joists. However, Douglas fir is so dense it needs to be incised along its surfaces to accept pressure treatment. Southern yellow pine is nearly as strong as Douglas fir and doesn't have to be incised, which is why it is so popular for pressure treatment. Since joist spans in a gazebo will probably be short, the strength of the framing lumber is not likely to be a big factor in your choice. In most cases, it's most cost-effective to build the hidden floor frame from whatever pressure-treated wood is available in your area.

using pressure-treated wood safely

The most important precaution when working with pressure-treated wood is to wear a dust mask when cutting it. In fact, it's a good idea to wear a dust mask when cutting any wood. Remember to wear safety glasses or goggles when working wood with any power tools. Never dispose of pressure-treated wood by burning, because it releases the chemicals into the air. Always dispose of these scraps in the trash.

pressure-treated wood

When you get right down to it, termites and fungi are looking for the same basic things as you and I. The little critters just want something to eat, water to drink, and a warm place to live. Unfortunately, termites and some fungi have chosen wood as their favorite food. One way to keep the critters away is to poison their food supply by building with pressure-treated wood. In many situations this is a very practical solution, specifically when wood is used in or near the ground where it may remain wet or damp. In the case of a gazebo, that's likely to mean building the posts and floor framing from pressure-treated wood.

Pressure treatment consists of placing wood—one commonly used species is southern yellow pine—into a chamber that uses pressure to force chemicals into the wood. The latest generation of chemicals contains copper, which drives away the termites and most fungi. The copper is combined with an organic *co-biocide*—a mixture of organically-based chemicals that repels fungi that aren't affected by copper alone. Depending on the brand, the formulation can be alkaline copper quaternary or copper azole. In both cases the co-biocide is designed to replace the arsenate component in chromated copper arsenate (CCA), the chemical mix that was used for decades in pressure treatment. CCA was phased out for residential use during 2002.

If you use pressure-treated wood for posts in the ground, be sure to get wood that is rated for ground contact. This means that the treatment forces the chemicals deeper into the wood for more protection. For parts of your gazebo that will be at least 12 inches off the ground, you can use wood that is rated for above-ground use.

Pressure treatment makes wood very *rot* resistant, but it doesn't make the wood more *weather* resistant. Many people must be confused about this. How else do you explain the proliferation of decks made from pressure-treated lumber? When a piece of pine emerges from the pressure tank, it is still pine, a species that is more prone to cracking, splintering, and twisting in response to humidity changes than some other species, including redwood, cedar, and cypress. Some manufacturers of pressure-treated wood have addressed this by adding a water repellent to the treatment mix. While I don't have any direct experience with this product, I do know that it helps to regularly coat exposed treated wood with a deck preservative.

In choosing wood for your gazebo, it's important to understand the difference between sapwood and heartwood. Sapwood is the outer part of the tree whose cells transport sap and store food. As the tree grows and adds outside sapwood rings, the inside rings, or cells, become heartwood. Since the structure and density of the cells doesn't change, the heartwood of any given species has the same strength, stability, and weather-resistance as the sapwood. However, heartwood cells stop transporting sap and storing food. This change from sapwood to heartwood is accompanied by an increase in the production of substances called extractives. In many species, these extractives darken the heartwood. It is important to note that while sapwood is never rot resistant, in certain species, these extractives make the heartwood very rot resistant.

Woods that are most prized for outdoor building are lightweight species that produce rot-resistant heartwood. For lumber produced in North America, this most often means redwood or western red cedar. Other woods may be available regionally. For example, cypress, sometimes called bald cypress, is very rot resistant, and can be found in southern parts of the United States.

REDWOOD

Because redwood grows only on the California coast, it can be quite expensive if you don't live in that region. It's strong enough to use for framing, such as the floor joists of your gazebo, but it's unlikely you'll want to pay a premium price for hidden wood when pressure-treated lumber will do the same job. Redwood's sapwood is almost white, while the heartwood varies from a light cherry to a dark mahogany. Redwood has a beautiful straight grain. You may want to pay the higher price for it for visible parts of your gazebo such as flooring, rails, and posts.

If you plan to use a piece of redwood lumber near the ground, look for the word Heart in its grade stamp. This means the piece consists of all heartwood. The all-heartwood grades range from Construc-

redwood lumber grades

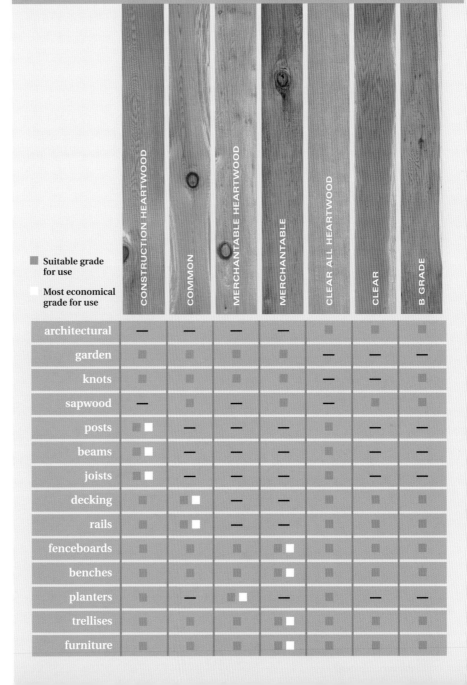

Legend:
- ■ Suitable grade for use
- □ Most economical grade for use

	CONSTRUCTION HEARTWOOD	COMMON	MERCHANTABLE HEARTWOOD	MERCHANTABLE	CLEAR ALL HEARTWOOD	CLEAR	B GRADE
architectural	—	—	—	—	■	■	■
garden	■	■	■	■	—	—	—
knots	■	■	■	■	—	—	—
sapwood	—	■	—	■	—	■	■
posts	■ □	—	—	—	■	—	—
beams	■ □	—	—	—	■	—	—
joists	■ □	—	—	—	■		
decking	■	■ □	—	—	■	■	■
rails	■	■ □	—	—	■	■	■
fenceboards	■	■	■	■ □	■	■	■
benches	■	■	■	■ □	■	■	■
planters		—	■ □	—	■	—	—
trellises	■	■	■	■ □	■	■	■
furniture	■	■	■	■ □	■	■	■

tion Heart, which contains knots, to Clear All Heart, which is knot-free and quite expensive. If the lumber stamp just says Clear, it only means there are no knots. This makes it a nice-looking piece of wood, but it is not suitable for use near the ground. The chart on page 33 is a guide to redwood grades and their use.

RED CEDAR

To avoid confusion when discussing red cedar, note that there are two types: *eastern* red cedar and *western* red cedar. Eastern red cedar, which grows throughout most of the eastern half of the United States, has heartwood that is extremely rot resistant. Unfortunately, it doesn't grow to diameters big enough for making lumber and is mostly used as fence posts. How-

ever, eastern red cedar has always been the wood of choice for builders of rustic garden furniture and structures, including gazebos. For our purposes, when we talk about cedar from now on, it will be in reference to western red cedar.

Red cedar has sapwood that is nearly white, with heartwood that is bright or dull red. Cedar is less dense than redwood, making cedar more stable and weather resistant than redwood but not as strong. As a result, carpenters don't commonly use cedar for structural members, such as floor joists, making large-dimension pieces hard to find at the local lumberyard. However, because of its stability in the face of moisture changes, cedar is popular for use as decking, with 2 x 4s, 2 x 6s, and 5/4 x 6s readily available. You should also be able

to find red cedar posts in nominal 4 x 4 dimensions.

Keep in mind that all wood is much stronger when used in compression (with the grain running vertically, as in a post) than it is in tension (when used horizontally). This means cedar posts will be more than strong enough to support your gazebo roof if you choose to use them. Cedar shingles and shakes, discussed a little later, are also made of western red cedar.

Because cedar is used primarily for decking, where appearance is important, its grading system emphasizes whether it is clear or knotty. There is, however, a grading category called Clear All Heart, which subdivides all heartwood lumber into various grades, based on numbers of knots and other defects. As with redwood, look for All Heart lumber for use near the ground.

Eastern red cedar doesn't grow large enough for making lumber, but its extreme rot resistance makes it attractive to rustic builders, including Romancing the Woods of Woodstock, New York, creators of this gazebo. PHOTO COURTESY OF ROMANCING THE WOODS.

MAHOGANY FLOORING

For many people, the word mahogany conjures images of beautiful, deep-red hardwood used to make fine furniture. If using mahogany for your gazebo floor seems extravagant, rest assured you won't be getting that fancy stuff, sometimes called true mahogany or American mahogany, when you saunter into your local home center. What you will get is mahogany porch flooring, which can consist of any of about 200 species of wood that comes mostly from the Philippines or sometimes from Africa. You may have heard of luaun mahogany, which is one species from the Philippines.

When applied to porch flooring, the term mahogany is really more of a grade than a species designation. While all mahogany porch flooring makes a flooring that is stable (like true mahogany), it is not necessarily very rot resistant. The wood is milled into tongue-and-groove 1x boards that are actually ¾ inches thick, or 5/4 boards that are approximately 1 inch thick. Because the wood is stable, the tongue-and-groove joints won't buckle in humid weather or open up in dry periods.

Mahogany flooring wears well, especially under a few coats of outdoor floor paint. Due to the mixture of species, there is a wide color variation from board to board, which can look interesting when varnished. Overall, mahogany flooring is a great choice for use outdoors, as long as a roof protects it from prolonged soaking. For a traditional Victorian-style gazebo, it looks much more appropriate than decking boards.

prettier posts

IN many gazebo designs, the posts are set in the ground or on concrete piers close to the ground. The posts rise through the floor to support railings and the roof. If you don't like the look of pressure-treated wood, or any framing lumber for that matter, you can wrap the posts from the floor up with boards of a more attractive wood.

Beneath this dressed-up post hides a lowly 4 x 4. The 4 x 4 is clad with 1x pine, available at any lumberyard. A capital of 2x lumber with cove molding underneath creates a launching pad for classic Carpenter Gothic brackets. PHOTO BY TERRY WILD.

using decking boards

When deciding on the type of wood for the floor of your gazebo, consider using 5/4 (five quarter) decking boards. These boards, available in pressure-treated wood, redwood, and western red cedar, are approximately 1 inch thick and 5½ inches wide. The two top edges of each board are rounded for appearance and to help prevent splinters. In many designs, but especially for smaller gazebos, a floor of 5/4 lumber will look less clunky than those made of 2 x 4s or 2 x 6s.

Wood Finishes

You could, of course, use no finish at all on your gazebo. Any wood left to the weather will turn gray within a year or so. The gray color is actually a thin layer of partially deteriorated wood fibers. Further weathering causes loss of these fibers, but the process is so slow—approximately ¼ inch per century—that you don't need to worry about it. As long as wood can dry before fungi take up residence, rot isn't a problem. Vertical boards that allow water to run right off fare better than horizontal boards, and sunny spots are better than shady spots. Mildew is another problem that can attack damp, unfinished wood. While mildew doesn't rot the wood, it's unsightly.

One of the functions of wood finish is to partially seal the cut ends of the wood cells to inhibit expansion and contraction caused by moisture moving in and out of the wood. The expansion and contraction causes cracking, twisting, splintering, and cupping. As we discussed, lighter-weight woods are less susceptible to these problems, making an unfinished cedar railing fare better than a railing of unfinished, pressure-treated, southern yellow pine.

Varnish is not generally recommended for outdoor structures. It forms an inflexible film that cracks when wood expands and contracts, causing the finish to peel when water gets under it. However, there is one exception. You may consider using a good marine varnish on mahogany decking boards that are protected from the weather.

For most parts of your gazebo, you have four finishing choices:
- Do nothing
- Apply a preservative
- Use a stain
- Paint

Surrounded by a riot of daffodils and tulips, this gazebo doesn't need to contribute any color to the landscape. Therefore, no finish was applied, allowing the gazebo to gracefully turn to gray. PHOTO BY ALAN AND LINDA DETRICK.

wood preservatives

Use wood preservatives when you want to maintain the natural color of the wood. The least expensive formulations are basically paraffin or some other wax dissolved in mineral spirits. They don't do much to retain the wood's natural hue. As the wax soaks in to clog the cells, the mineral spirits evaporate. You can use these formulas on pressure-treated wood when you want to lose the green tint and want the wood to turn gray. For more money, you can buy a preservative formulated to block ultraviolet (UV) light. It's the UV rays that destroy the surface fibers and turn wood gray. Always use UV-blocking formulas when you want to preserve the color of redwood or cedar. Other additives found in wood preservatives include fungicides and mildewcides—always a good idea to use in damp locations.

paint

Like varnish, paint forms a film on the surface of the wood rather than penetrating the cells. But paint is more flexible than varnish, allowing it to stretch and contract with wood movement. Paint also blocks the UV rays that deteriorate surface fibers. This is an important characteristic of paint since it won't adhere to gray, weathered wood fibers.

Most paints today are either alkyd oil based or water-based latex. Traditionally, the durability of oil-based paints made them preferable. But recently, because of concerns about volatile organic chemicals (VOCs), paint manufacturers have put a lot of resources into improving latex paints. Besides greatly reducing the VOCs released into the air, the big advantage of latex paints is that you can clean tools, brushes, and your skin with water. Oil-based paints require mineral spirits for cleanup. In my experience, the latest generation of latex paints is so good that I see no reason to go through the expense and bother of using the oil-based versions. There is one possible exception: The Western Red Cedar Association recommends using an alkyd-based primer on cedar, although they acknowledge that there are latex primers designed to go over cedar.

The most important point to remember about primers is to use one. Primer is essentially the glue that holds paint to new wood. If you skip the primer, the paint will surely peel. Make sure to use enough primer to obscure the grain. Then give the wood two coats of exterior paint.

Floor paint, one type of specialty paint, is worth your attention. As the name implies, this type of paint is made specifically to stand up to foot traffic. All the floor paint I have seen is specified for interior or exterior use. Always check the label to make sure before choosing one designed for your outdoor structure. A good floor paint will contain epoxy that makes the surface really tough and long-lasting.

stains

Stains for outdoor use can be either semi-transparent or a solid color. Both have an oil or resin base that penetrates into the wood, much like a preservative. The colors come from pigments that stay on the surface as the stain soaks in. Stains are available in a wide range of colors.

Semi-transparent stains, as you would expect, contain less pigment so that the grain of the wood isn't hidden. Solid stains leave enough pigment to obscure the grain without completely obscuring the texture of the wood. With solid-color stains, the oil or resin doesn't penetrate as deeply into the wood as it does with semi-transparent stains.

I find that stains, especially the semi-transparent variety, look best on woods with open, straight grain, such as cedar. Smooth pine trim boards sometimes absorb the stain unevenly, which can look blotchy. You can help prevent this problem by pretreating the lumber with a paintable wood preservative. In my opinion though, the grain pattern on most pine you'll find in lumberyards today is not distinctive enough to highlight with stain anyway.

Because they don't soak as deeply into the wood, solid-color stains act like thinned paint. Solid stains are not recommended for floors where the surface pigment would quickly wear away.

Because stains are absorbed into the wood, they won't eventually peel like paint. Also, unlike paint, adding additional coats of stain won't cause a buildup that eventually must be removed. The trade-off is that stain is less durable than paint and must be recoated more often.

Roofing

I've seen gazebos roofed with everything from clay tiles to metal or slate—even thatch. Since gazebo roofs are generally small, price is usually not a deciding factor in choosing roofing. A more important consideration is the fact that a gazebo often has many converging hips—the lines where roof planes meet at an angle. This means that, unless you're going to have special roofing panels or pieces fabricated, you'll want roofing material that looks good and is easy to cut and install with common carpentry tools.

Often, gazebos have roofs that aren't designed to keep out the rain. You might prefer a roof of lattice or slats to admit dappled sunlight when the gazebo is in use. But if you do want a waterproof roof that you can install yourself, the most likely choices are cedar or asphalt shingles.

A roof of spaced slats diffuses direct sunlight and gives this small garden gazebo a light and airy feel. PHOTO BY MARK TURNER.

The custom-fabricated metal roof, with its tall finial, makes this small gazebo an elegant and intimate garden destination. PHOTO BY DENCY KANE.

The exposed shingle bottoms contrast nicely with the color of the skip sheathing and rafters, and display the gazebo's roof framing. PHOTO BY INGELA NOREN.

cedar shakes or shingles

For most gazebos, a roof of cedar shakes or shingles is the best choice. Shingles are smoother, while shakes have a more rustic look. Both are great for adding a natural look to a garden or woodland setting. They can be installed on *skip sheathing*, which is 1 x 4 or 1 x 6 boards with spaces in between. This installation looks particularly nice from inside the gazebo.

A *shingle* is sawn from a solid block of wood, while a *shake* has at least one face that has been split. Both products come in lengths of 18 or 24 inches, while shingles are also available in 16-inch lengths. The length you choose depends on the exposure—how much of the shingle you leave exposed to the weather, as opposed to being covered by shingles above. The steeper your roof, the greater you can make the exposure. Shingles and shakes are both available either tapered or uniform along their length. In general, shakes are thicker than shingles, especially the thicker end called the *butt*.

The difference between a *split shake* and a *sawn shingle* is not just aesthetic. When you split wood, the pieces will separate along the cells, leaving them intact. When you saw wood, you sever the cells, which structurally weakens the wood. For roof applications this means that water can wick in through the severed cells. This isn't the end of the world, since properly installed shingles will dry out before rot can take hold. But the fact remains that shakes are more durable, especially on lower-pitched roofs, which dry more slowly.

I prefer using *handsplit shakes*, which have split faces and sawn backs. The result is maximum durability to the weather and a smoother look inside. If you do choose cedar shingles, you might want to consider adding a couple of courses of *fancy butt shingles* to your gazebo roof. The butts of these shingles come in various shapes, and their complex profiles are a great way to add some flair to your gazebo.

While both have weathered to a similar silver gray, the cedar shakes on the gazebo on the left add texture to the expansive roof, while the cedar shingles on the right look more refined. PHOTOS BY DICK DIETRICH.

asphalt shingles

Asphalt shingles are another roofing choice for gazebos. They are quite inexpensive, especially when you consider they're designed to last for a few decades when installed on a house. You should get even more mileage using them for a gazebo roof, since it's not subjected to the same heat extremes as a typical house roof built over an attic. In addition, asphalt shingles come in many colors and styles, they easily conform to any curves you will include in the roof, and they are very easy to install.

For my own preference, I find that the granules-embedded-in-tar surface of asphalt shingles looks better from a distance on a house rather than on the close-up view you get with the typical gazebo. Also, to my eye, asphalt shingles seem disproportionately large, especially on a small gazebo. Of course, this is my opinion, and you will have to decide for yourself.

Another consideration when deciding to use asphalt shingles is that you must install them on plywood sheathing. Unless you install a ceiling in your gazebo, the underside of the plywood will be exposed. Of course, exposed plywood can be attractive. You can use a better grade of plywood and paint it—perhaps choosing a color that contrasts with the rafters. Another solution is to use T-1-11 plywood siding. These plywood sheets have an attractive rough-sawn face incised with grooves to look like individual boards. Used upside down as gazebo roof sheathing, T-1-11 makes a nice ceiling treatment.

This bell-shaped gazebo roof would have been difficult to cover with anything other than flexible asphalt shingles. PHOTO BY TERRY WILD.

Fasteners

To assemble your gazebo, you'll need a variety of nails, wood screws, and metal framing connectors, such as joist hangers and post anchors. Depending on the design, you may also need heavy-duty fasteners, such as lag screws and carriage bolts. Later in the book, we'll get specific about where and how to use various fasteners as well as the sizes you'll need as I take you through the gazebo construction. For now, let's look at the general characteristics of various fasteners.

nails

For any kind of outdoor construction, you'll want to use *hot-dipped galvanized nails*. These are nails that have been dipped in hot zinc. The zinc coating greatly delays the inevitable rusting that causes nails to stain wood. You may also find nails that have been galvanized by electroplating instead of hot dipping. Don't bother with these since their coating is much less effective. In fact, the California Redwood Association and the Western Red Cedar Association both recommend hot-dipped nails as the fastener of choice. It's easy to tell the difference; the surface of the hot-dipped nail is much rougher than the electroplated nail. As a bonus, the rough surface of the hot-dipped nail also improves the nails' holding power.

Most of your framing work will be done with *common nails*. These nails have thick shanks and broader heads, characteristics that increase the nails' holding power and make them easy to drive without bending. For exposed trimwork pieces, such as fascia boards, that just need to stay up rather than serve as support for other pieces, use *finish nails* or *casing nails*. These nails

have smaller heads and are much thinner than common nails. They allow you to use a nailset to drive them below the surface, leaving an inconspicuously small hole. I usually use casing nails for outdoor applications. They have dull points that are less likely to split the wood while you are nailing or weaken the wood enough to cause a split later when the wood expands and contracts. All of these nails are available with a hot-dipped galvanized coating.

You'll also need some special-purpose nails. Galvanized *roofing nails* are short nails with flat heads for holding asphalt roofing in place. For cedar shakes or shingles, you'll use galvanized or stainless steel *box nails*, which have thin shanks to prevent splitting when working with thin, brittle wood. For blind-nailing mahogany flooring, use hot-dipped galvanized *spiral deck nails*. Their spiral shape prevents the nails from pulling up, which causes squeaks when the wood expands and contracts with temperature changes. (See Blind-nailing Tongue-and-Groove Flooring, page 69.)

screws and bolts

If you choose to use lumber decking instead of tongue-and-groove flooring, you'll most likely attach it to the joists with galvanized, bugle-head screws, often sold as *deck screws*. These screws are similar to the familiar black drywall screws, which also have bugle-shaped heads. The difference is that deck screws have a coarse thread and a sharp point for power driving into softwood without predrilling. These screws are still most commonly found with Phillips heads, but square-drive screws are becoming increasingly popular because the driver engages the screw more effectively, reducing the likelihood that the screw head will become stripped. Often, square-drive screws accept a Phillips head as well. These combination-head screws are available with a proprietary green weather coating that makes the screws less conspicuous in pressure-treated wood.

Depending on the design of your gazebo's frame, you may need

FROM LEFT TO RIGHT: carriage bolt with washer and nut, lag screw, J-bolt

lag screws or *carriage bolts*. These heavy-duty fasteners are perfect for load-bearing situations such as attaching a beam to the side of a post. Lag screws are essentially oversize screws with hex heads that you drive with a wrench. Carriage bolts make an even stronger connection because they go completely through both framing members and are fastened with a washer and nut on one end. The head on a carriage bolt is rounded with a squared-off shoulder underneath. The squared shoulder engages the hole as you tighten the nut. The rounded head has a smooth finished appearance for applications where the head will become a visible part of your construction.

For anchoring post bases or standoffs to concrete, you'll use a *J-bolt*, so named because one end is turned up in the shape of the letter. The other end is threaded. To position it, you wiggle the J-shaped end into freshly poured concrete to seat it. The opposite threaded end will extend from the concrete to accept a nut for fastening the base or standoff.

metal framing connectors

These days there are steel connectors designed for virtually every situation in which two pieces of wood meet. For the most part, these fasteners allow minimally skilled workers to quickly make strong connections on production framing jobs. You, on the other hand, are not building 400 condominium units. You are crafting a unique and graceful garden structure. Although there are connectors you can use for visible areas, such as where railings join posts, I'm going to assume you won't want to use metal connectors where they will be visible.

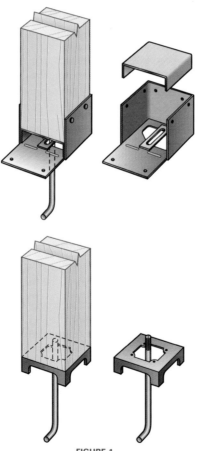

FIGURE 1
Post Bases and Standoffs

POST BASES AND STANDOFFS

If your posts will be supported by concrete piers, you'll need to attach them to a *post base* or *post standoff*, as shown in figure 1, to hold the posts in place and to keep their bottoms off the concrete where they would wick moisture. The strongest post bases are made from one piece of steel, and have prongs that you set into wet concrete. This strength is necessary for structures like patio roofs and carports with no lateral support between the floor and roof. Since you don't need lateral support for gazebo posts that will be locked in place by the floor frame, it's best to choose the style of post base that attaches to a J-bolt set in the concrete. These bases let you adjust the position and angle of the posts when you install them, making the job much easier.

A post standoff is a more attractive alternative for gazebos with patio floors where the bottom of the post is visible. Rather than nailing through metal tabs on the outside faces of the posts, you drill a hole in the bottom of the post, then insert the post onto the J-bolt. Keep in mind that you can't use standoffs or adjustable bases for structures where there's no support between the floor and the roof, such as those that don't have a floor frame or at least a railing. In the strictest definition, that kind of structure is really a pavilion and not a gazebo.

POST CAPS

In many gazebo designs, posts run from the piers to the roof, eliminating the need for post caps. In other designs, short posts end underneath joists or beams, necessitating the use of post caps to make these connections. You can find caps to fit 4 x 4 posts, with a single 2x joist or doubled 2xs, as shown in figure 2. You can also find caps to fit 6 x 6 posts with doubled or tripled 2x beams.

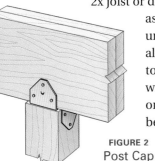

FIGURE 2
Post Cap

JOIST HANGERS

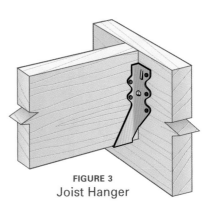

For most designs, using joist hangers makes a lot of sense. Because they are hidden under the deck, aesthetics is not an issue. Hangers make it easy to locate joists exactly where you want them while ensuring that these critical load-bearing connections are secure (see figure 3). They come sized to fit every dimension of lumber, including double-width to accommodate doubled beams or joists. If your design includes joists that meet at an angle, you can find hangers that you can skew to an angle up to 45°. You can also find hangers that can be angled down as well as skewed to the side for connecting jack rafters into the side of main hip rafters. However, I wouldn't use these on anything but rafters that you intend to hide.

FIGURE 3
Joist Hanger

CONNECTORS FOR SIX-SIDED GAZEBOS

There are three types of connectors designed specifically for six-sided gazebos as shown in figure 4. The one I wouldn't hesitate to use is a *joist/post tie*. This piece of hardware makes it easy to attach rim joists to posts at the proper angle, avoiding some tricky toenailing (nailing at an angle). In almost all designs, trim boards will hide the joist/post ties.

The second type of connector attaches rafters to plates, and is called a *rafter/plate tie*. This ties the rafters to the post as well as to the plate to make a strong connection. It avoids toenailing, and saves you from cutting a bird's-mouth in the bottom of each rafter. However, these connectors aren't really necessary, and I don't recommend them unless your gazebo design will hide them. On page 129, I will show you how to cut a bird's-mouth.

The third connector is a *top rafter tie*. This fastener joins all six rafters at the top by sandwiching them between two pieces of formed steel. Personally, I don't want to see a chunk of steel at the pinnacle of my gazebo ceiling, so I wouldn't use this tie unless I planned to cover the bottom of the rafters.

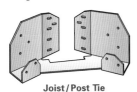

Joist/Post Tie

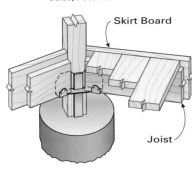

Skirt Board

Joist

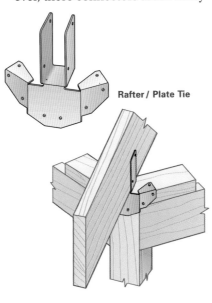

Rafter / Plate Tie

FIGURE 4
Connectors for Six-sided Gazebos

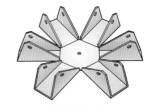

Top Rafter Tie

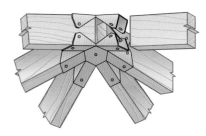

CHAPTER FOUR

Starting Out on the Right Footing

4

IF you want a wooden deck for your gazebo, there are two building approaches for anchoring it to the ground. You can support the structure on poured concrete footings, or you can sink posts directly into the ground. A third choice is to not anchor the gazebo at all, but to set it directly on the ground, perhaps on a bed of gravel, or raise it just a bit using precast concrete piers set on the ground.

If you don't want a wooden deck, there are several alternatives from which to choose. You can build your gazebo on top of a concrete pad. Or perhaps you want a brick walkway to flow into the gazebo's floor made of brick dry laid in sand. This is a great way to unify the look of your yard.

Each method has its pros and cons depending mostly on your site. In this chapter, I'll give you the information you need to choose a method, then show you how to accomplish each one.

Even in winter, your gazebo can be an attractive focal point. Knowing your climate is essential in planning a footing.
PHOTO COURTESY, VIXEN HILL.

Laying Out the Corners

Regardless of the foundation you choose, the first step is locating the exact position of each corner of your gazebo. This means finding the exact center of each post in most designs, or finding the outside corners of rim joists in some designs, though in both cases the layout method is the same. In this section I'll start by showing you how to lay out the corners for gazebos with four, eight, or six sides. By the way, you can use these same methods to lay out the corners of a concrete or brick base even though it may not be the same size as your gazebo.

laying out a four-sided gazebo

To lay out a rectangle, we'll pay a visit to our old high school buddy Pythagoras. Remember him? He's the guy in geometry class who figured out that for a right triangle, the hypotenuse squared is equal to the sum of the squares of the other sides. For our purposes the Pythagorean theorem translates to the 3–4–5 method, a simple way of checking that a corner is square. If one side of a right triangle is 3 feet long and the perpendicular side is 4 feet long, then the hypotenuse will be 5 feet long. Of course, you can use any multiples that are convenient. Remember that longer measurements will increase the accuracy of the method. For a 12-foot-wide gazebo, using 6, 8, and 10 feet in the method will work well.

1 **Locate the corners.** In this step, use rough estimates for placing the corners. Begin by placing a stake in the ground at the location of one corner. Next, use a tape measure to locate the three remaining corners. Then drive stakes in each location.

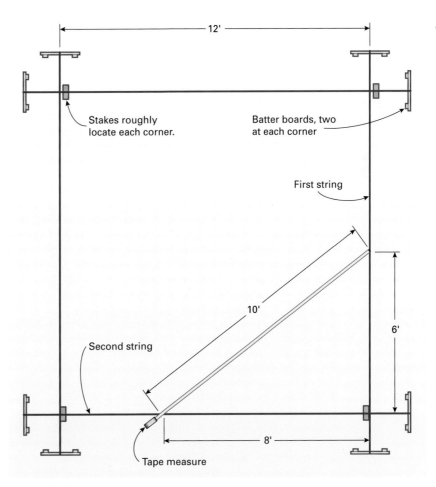

Stakes roughly locate each corner.

Batter boards, two at each corner

First string

Second string

Tape measure

FIGURE 1

To square a corner, start with one end of the tape at the 8-ft. mark on one string, then adjust the other string until the 10-ft. mark on the tape aligns with the 6-ft. mark.

Use a level or a plumb bob to transfer the corner locations to the ground. Flag the location with a nail poked through a bit of bright cloth or plastic.

2 **Set up batterboards.** Erect two batterboards at each corner, placing one at each side of the corner approximately 2 feet from the stake.

3 **Set the strings.** First, tie a string across two opposing batterboards as shown in figure 1. Adjust the string until it's positioned where you want it for one side of your gazebo. Next, tie another string to a third batterboard. Have a helper stretch the string until it is taut, allowing it to cross the first string to form a corner. Measuring from the point where the strings cross, make a mark at 6 feet on the first string and a mark at 8 feet on the one your helper is holding. Hook your tape measure on the 6-foot mark and stretch it to the other mark. Have your helper

adjust the string until the mark aligns with 10 feet on your tape measure. Tie the string down. You have just used the 3–4–5 method to create a square corner. Do the same for the three remaining corners.

4 **Check your layout.** Check that all the sides are the length you want. Measure across both diagonals. If the measurements are the same, the corners will be square.

5 **Flag the corners.** Poke a nail through a brightly colored piece of cloth or plastic. Use a plumb bob or a level to locate the point directly under the points where the strings cross. Place a nail in the ground at that point. Drive a nail into each batterboard at the point where you've tied the string. Then remove the strings.

what's a batterboard?

A batterboard is simply two stakes set in the ground approximately 2 feet apart with a horizontal board nailed or screwed between them. After roughly locating the corners of your gazebo, you'll set up lopposing batterboards approximately 2 feet outside each corner, making eight batterboards for a four-sided gazebo. To delineate each side of the gazebo, tie string between the crosspieces of each opposing pair of batterboards. Mason's twine is best because it won't sag.

Next, adjust the strings until they cross at the exact

points where you want to locate the center of your posts, the outside of your rim joists, or your concrete forms. Transfer this point to the ground and mark it with a nail poked through a piece of cloth or colored plastic.

Hammer a nail into the top of each batterboard at the exact point where you tied the string. Now remove the strings to allow you to either dig holes for posts or piers, or to excavate for a concrete pour. Later, when you want to locate these exact points again for setting post anchors, posts, or erect forms, all you need to do is tie string between the batterboard nails.

Once you establish square corners, hammer a nail into the top of each batterboard to record the string positions.

To lay out an eight-sided gazebo, start by using batterboards and a string to lay out a square just as you would for a four-sided gazebo. Next, lay out the diagonals with stakes and two more pieces of string. This, along with a tape measure, is all you need to find the eight corners.

1 **Lay out the square.** Let's say, for example, that you want 12 feet between opposing posts. First lay out a square, making each side 12 feet long. Mark the mid-point of each side, in this case 6 feet from each corner. These are the locations of four of the post centers, or corners. Flag these in the ground by poking a nail through a piece of cloth or plastic.

2 **Lay out the diagonals.** Drive a stake into the ground at each corner of the square. Tie strings between these stakes to delineate the diagonals. Where diagonals cross, drive another stake directly under them. Drive it far enough to allow the strings to graze its top. Mark the intersection of the strings on the stake.

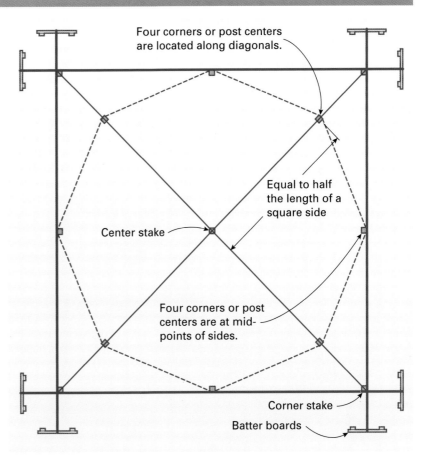

Four corners or post centers are located along diagonals.

Equal to half the length of a square side

Center stake

Four corners or post centers are at mid-points of sides.

Corner stake

Batter boards

FIGURE 2

X marks the spot. Crossed diagonal strings locate the exact center of the gazebo. Mark that point on the center stake.

3 **Locate the remaining corners.** Measure along each diagonal from the midpoint marked on the center stake to a distance equal to half the length of the square sides, as shown in figure 2. Again, in our example, we would measure 6 feet. Measure out in all four directions to find the locations of the remaining corners, or post centers. Flag these corners in the ground using nails that have cloth or plastic poked through them. Remove the strings when you dig the piers. Leave the batterboards, corner stakes, and the center stake in place to help you locate the center of the piers later. Some designs call for a pier at the center location. In that case, you'll need to remove the center stake when you dig pier holes.

laying out a six-sided gazebo

If you draw a circle of any size, and mark six equidistant points along the circumference, the distance between those points will be the same as the circle's radius. This is the simple secret to laying out a six-sided gazebo. To take advantage of this convenient mathematical fact, we'll abandon the batterboards and string. Instead, we'll lay out our six corners with a giant compass made from a 2 x 4 and a couple of nails.

1 **Make the compass.** Let's say you want the distance between the center of opposing posts to be 12 feet. This means the radius of your compass will need to be 6 feet. To do this, take an 8-foot-long 2 x 4 and make a mark approximately 1 foot from one end, centering the mark on the board's face. Make another mark exactly 6 feet from the first. Drill a ¼-inch pivot hole through one mark. Then drive a nail all the way through the 2 x 4 at the other mark.

2 **Scribe the circle.** Locate the center point of your gazebo. Drive a short 2 x 4 stake into the ground until its top is almost flush with the ground. Place a nail in the pivot hole of the compass and drive the nail into the top of the stake. Now use the nail on the other end of the compass to scribe a circle. To be able to see the circle, fill the little furrow you make with lime or sand as you go. You can make a handy lime or sand dispenser by ripping a hole in one bottom corner of a small plastic bag with a zipper-like closure.

3 **Lay out the posts.** Remove the compass from the center pivot stake. Pick a point along the scribed circle for the first post. Place the nail of the compass in the ground at this point. Swing the compass until the nail pivot hole intersects the circle. This is the location of your second post. Lay out the third

A 2 x 4 compass, a center stake, a couple of nails, and a bag of lime are all you need to lay out the perimeter of a six-sided gazebo.

post from the second, and so on. Poke a nail through a brightly colored piece of cloth or plastic, and use it to mark each location. Leave the center pivot stake in place and keep your compass. After you pour the piers, you'll use the stake and compass to locate the exact center of each pier as shown in figure 3.

With one of the compass's nails positioned in the location for the first pier, locate the second pier by swinging the second nail until it intersects the circle.

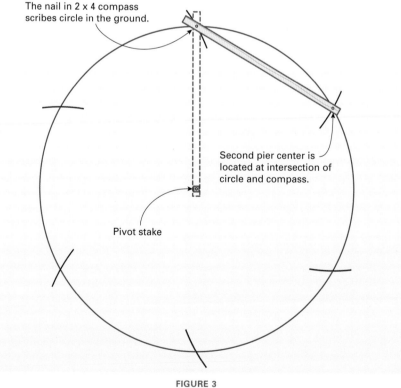

The nail in 2 x 4 compass scribes circle in the ground.

Second pier center is located at intersection of circle and compass.

Pivot stake

FIGURE 3

Choosing a Foundation

When choosing the appropriate foundation for your gazebo, the first question you should ask is whether the ground freezes in your climate. If it does, it means that the ground may heave in the wintertime as moisture in the soil freezes and expands. To create a truly solid and stable gazebo, you need to isolate it from that movement by providing a footing that extends below your local frost line, which is the depth where the soil provides enough insulation to prevent moisture from freezing. Depending on where you live, this can be anywhere from 1 to 4 feet.

Another approach is to set your gazebo on a 4- to 6-inch bed of gravel. Because it drains better than soil, the gravel itself won't suffer from frost heave. The gravel also absorbs the movement from frost below it. After a rain, the excellent drainage from gravel guarantees that the gazebo's wooden structure won't be sitting in water, which is damaging to any wood, including pressure-treated lumber.

If the ground doesn't freeze in your climate, you won't need a sunken footing for a small structure such as a gazebo. In this case, your only concerns are providing a level base and keeping the bottom of your gazebo out of standing water. If heavy rains are common where you live, setting your gazebo on a bed of gravel that is a couple of inches thick should do the trick. You can also keep the bottom of the gazebo dry by raising it on precast concrete piers that you set directly on the ground.

do I need a permit?

When in doubt, it's always a good idea to stop by your town's building department to see if you need a building permit for your gazebo. Usually you'll need one if you are sinking piers or posts or pouring a concrete slab because the gazebo will be considered a permanent structure. If the gazebo will simply be set on the ground, chances are you won't need a permit.

poured concrete footings

This is my favorite method for locations where the ground freezes. The piers themselves are easy to make, thanks to round cardboard forms. These forms are the overgrown cousins of paper towel tubes, and you can get them at any home center or masonry supply store. After pouring the concrete into the form, place a threaded J-bolt in the wet concrete. When the concrete sets, fasten a metal post base to the J-bolt. This creates a base with two features I really like. First, it raises the bottom of the post off the pier so the post bottom never gets wet—water wicking in from the bottom is the biggest cause of rotting posts. Second, the bases can be adjusted approximately ½ inch in any direction, allowing you to get the posts exactly right even if you were unable to get the J-bolts in exactly the right position in the piers. Note that there is one situation in which you shouldn't use piers: If the gazebo won't have railings to provide lateral support to the posts, you should always sink the posts into the ground.

There are three ways to get concrete. For small quantities you can purchase prepackaged concrete in 40-, 60-, or 80-pound bags. To prepare it, you just dump the contents in a wheelbarrow and mix with water following the manufacturer's instructions. This is a great solution for filling concrete around a few posts, but quickly gets expensive for making piers—a single 3-foot-deep pier would take approximately three 80-pound bags.

The second option is to buy the ingredients—concrete consists of sand, gravel, Portland cement, and water. You can mix the ingredients by hand in a wheelbarrow, or you can rent a gas- or electric-powered mixer. If you mix it yourself, the proportions are 1 part cement, 2¼ parts sand, 1½ parts gravel, and approximately ½ part water. Mix

For the quantity of concrete you'll typically need for six or eight gazebo piers, it usually makes most sense to mix your own. Have gravel and sand delivered to the site, along with bags of Portland cement.

the dry ingredients first on a sheet of plywood or in a wheelbarrow. Then add the water and mix with a mason's hoe until all the ingredients are wet.

The easiest way to get concrete is to order it premixed from a concrete company that will deliver it in a truck. Since a ready-mix truck carries a lot more concrete than you'll need for gazebo piers, you may not meet the company's minimum delivery requirements. If you're willing to wait, they may agree to piggyback your order onto a larger order in your area if they have room in the truck.

If you're lucky, the ready-mix truck will have enough access to

your site to pour concrete directly from its chute into your forms. But more than likely, given the nature of placing gazebos in tucked-away spaces, you'll need to shuttle the concrete with wheelbarrows. It's a good idea to have at least two workers with two wheelbarrows on hand when the truck comes. However, if you have to wheel the concrete more than a few yards, I'd recommend not using a ready-mix truck. In this case, it's easier to mix your own concrete closer to your site.

Once you've decided how you will obtain your concrete, there are several steps to making poured concrete footings.

You can rent a concrete mixer, or you can save some money by using a mason's hoe or a shovel to blend the ingredients. Gradually add water to the dry ingredients.

calculating concrete

Ready-mix concrete is sold by the cubic yard. To calculate how much concrete you'll need for a pier, use the following formula: 0.7854 times the form diameter (in inches) squared times the depth of the form (in inches). This will give you the volume in cubic inches. For volume in cubic feet, divide by 1728. A cubic yard equals 27 cubic feet. As an example, here's how the formula works using a 3-foot-deep pier with a diameter of 8 inches:

8 inches squared = 64 inches
.7854 x 64 x 36 = 1809.56 cubic inches
1809.56 ÷ 1728 = 1.07 cubic feet
1.07 ÷ 27 = .04 cubic yards per pier

If you'd rather use the Internet than a calculator, search for "concrete calculator." You'll quickly find a place to plug in your numbers to get the amount you'll need.

If your ground is soft and you like to work, a clamshell digger is an effective way to dig your post or pier holes.

The sweat-free way to dig holes for posts or piers is to hire a fencing company and have them use a power auger attached to a truck, or in this case, a mini-excavator.

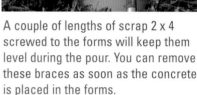

A couple of lengths of scrap 2 x 4 screwed to the forms will keep them level during the pour. You can remove these braces as soon as the concrete is placed in the forms.

1 **Dig the holes.** Your local building code will have requirements for the diameter and depth of your piers. Dig the holes wide enough to accommodate the forms you'll use and deep enough to reach below the frost line. If your soil is very soft, you can dig the holes with a clamshell digger. In most cases it's a good idea to rent and use a power auger, since the dimensions of the hole will be more precise when dug this way. If your soil is very rocky or has lots of roots, it can be tough to handle a one-man auger. In this situation it is better to rent a two-man auger and enlist a friend to help. If you don't want to pursue the do-it-yourself option, you can hire a fencing company to dig the holes.

2 **Set and fill the forms.** Begin by shoveling a few inches of gravel into the bottom of the hole. Place a length of the tube form into the hole and mark it for a cut that will leave approximately 2 inches above ground. Cut the tube to this length with a handsaw.

Place the tube in the hole and pack dirt around the outside of the form. As you do this, place a level across the top of the form in several directions to make sure the form is level. You might find it helpful to drive screws through the sides of two 2 x 4s to anchor the forms in position. When the form is securely in the hole, fill each form to the top with concrete and strike it off level with a piece of 2 x 4. There's no reason to worry about making the piers level to each other—the posts will absorb any height differences.

give it a rap

A common problem when filling forms with concrete is that air gets trapped in the mix in the form of bubbles. These bubbles can seriously undermine the strength of the pier you're making, so it's wise to remove them by tapping on the sides of your forms with your hammer before the concrete sets.

3 **Re-establish pier centers and insert J-bolts.** If you are building a four-sided or eight-sided gazebo, tie strings between the positioning nails you drove into the tops of the batterboards. With an eight-sided structure, tie additional strings to the corner stakes to re-establish the diagonals. Now measure along the strings as you did before to find the exact points for inserting the J-bolts. (See Laying Out an Eight-Sided Gazebo, page 47, step 3.) After pouring the concrete, wiggle a 6-inch J-bolt into the concrete at each of these points. In most cases, you'll want to leave about ½ inch of the bolt protruding from the top of the concrete, but check the post bases you purchased to make sure the raised part of the base will fit over the bolt. Use a square to check that the bolt sits at 90° to the pier.

The procedure for a six-sided gazebo ensures that the bolts will be equidistant from each other

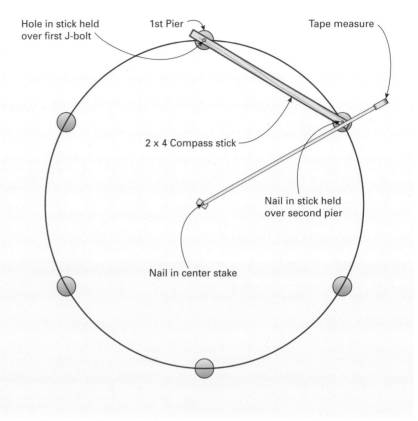

FIGURE 4

A mark on the string tells you exactly where to insert a J-bolt into the wet concrete.

even if the piers are a bit off. As shown in figure 4, first find the exact center of one pier and insert a J-bolt. Next, take your 2 x 4 compass and have a helper hold the hole in one end over the first J-bolt. Move the compass until the nail at the other end is at the center of the next pier. Hook your tape measure on the nail in the center stake and measure to the second pier. Adjust the compass and tape measure until the measurement over the nail matches the distance you want between pier centers. Insert a J-bolt into the concrete at this point. Continue moving the compass and measuring for each remaining pier.

Posts set directly in the ground can last for decades, as long as you use a rot-resistant wood and take steps to prevent water from wicking into the bottom of the posts. The dark center wood, or heartwood, of some species is extremely rot-resistant. Redwood, red cedar, and cypress are examples that are readily available in different parts of the country. (See Premier Outdoor Woods, page 33.)

However the most rot-resistant and economical wood to use for posts is pressure-treated wood. (See Pressure-Treated Wood, page 32.) Be sure to use wood rated for in-ground use only. Treated wood has an unnatural greenish color that you might find unsuitable for your gazebo. Keep in mind that the green fades to gray after a few years. And, if you decide to clad the posts, the color of the pressure-treated wood will not be an issue.

1 **Dig the hole.** Dig the hole to below your local frost line. The hole doesn't need to be any wider than the post you will use—in most cases the digger or auger you use will determine the width of the hole. If your gazebo won't have a floor or rails to increase lateral support, always make the holes at least three feet deep—this is essential, even if the ground doesn't freeze where you live.

2 **Prepare the hole.** Shovel a few inches of gravel into the hole. The gravel prevents water from wicking through the concrete into the bottom of the post. Tamp the gravel down using a length of 2 x 4.

3 **Leave posts long and loose.** Use posts that are at least 1 foot longer than their finished height. This allows a healthy margin for cutting the posts level once you finish the deck. For example, if you have holes that are three feet deep and you want the posts to end 10 feet above grade, use stock that is 14 feet long. Place the posts in the holes, but wait to fill in the dirt around them. This allows you to adjust their positions as you install the rim joists. Once you've adjusted the posts, fill around them with soil, tamping down after every few shovelfuls.

concrete in the hole?

You can fill around the posts with concrete instead of dirt, but for most gazebo projects there isn't a good reason to do so. Concrete won't protect the posts from moisture—in fact concrete is very porous and may actually wick moisture to the posts. The one good reason to add concrete around the posts is to increase the lateral support for gazebos that won't have a deck or rails to help tie the structure together.

To make concrete-filled holes, as shown in figure 5, dig them at least 6 inches below the frost line. Make the diameter of the hole two to three times the thickness of the posts. Put 2 to 4 inches of gravel in the hole, and tamp it down. Place your post in the hole and position it exactly where you want it. Plumb the post and brace it with lengths of 2 x 4s. Now you can start shoveling concrete into the hole. After every few shovelfuls, tamp the concrete with a 1 x 2 to work out any trapped air. For good drainage, slope the concrete away from the post at the top of the hole. After the concrete cures, you can cover it with soil if desired.

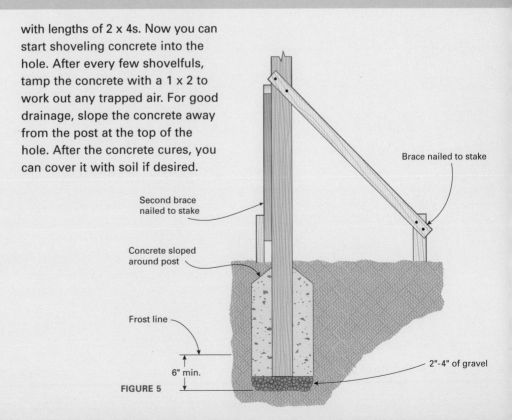

Brace nailed to stake

Second brace nailed to stake

Concrete sloped around post

Frost line

6" min.

2"-4" of gravel

FIGURE 5

dealing with brick or concrete floors

The brick surrounding this gazebo flows into the gazebo's floor to create a handsome, formal setting.
PHOTO COURTESY, VIXEN HILL.

Brick walks and patios are usually dry laid in sand, which is much easier to do than setting bricks in a poured-concrete base. Instead of mortar, the joints between bricks are filled with fine mason's sand. If drainage is a consideration at the site, laying several inches of gravel under the sand takes care of the problem.

If you would like to use dry-laid brick as the floor of your gazebo, you can still anchor the posts to piers, as shown in figure 6. To do this, pour the piers to make their tops level with the top of the sand bed.

Insert a ½- x 12-inch-long J-bolt into the wet concrete, leaving 7 inches of the bolt protruding from the concrete. Note that this next part is crucial: Use a level to make sure the J-bolt is perfectly plumb— this is more important than getting it perfectly square to the pier.

Next, lay the sand base and then the bricks. Drill a ⁹⁄₁₆-inch hole in the center of the bottom of the post. First make sure the hole will be square to the bottom of the base, then drill it approximately 7 inches deep. This will prevent the J-bolt from reaching the

bottom of the hole. Place a metal post end over the J-bolt, then insert the post over the J-bolt.

Many gazebo designs call for dressing up the post by cladding it with 1x (¾-inch-thick) boards. If you do this, keep the bottoms of the boards at least ½ inch off the brick. If you do clad the posts, it's a good idea to wrap the cladding with a baseboard. The horizontal grain of the baseboard won't wick up water, giving you the option of letting the baseboard sit on the brick. The result is a more attractive post. For additional embel-

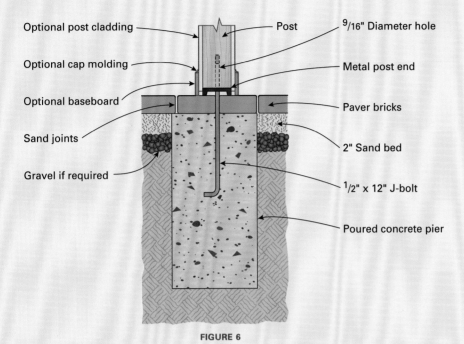

Optional post cladding — Post — 9/16" Diameter hole

Optional cap molding — Metal post end

Optional baseboard — Paver bricks

Sand joints — 2" Sand bed

Gravel if required — 1/2" x 12" J-bolt

Poured concrete pier

FIGURE 6

lishment, you can add cap molding on the tops of the baseboards.

You can anchor posts to a concrete floor in a similar way. Begin by setting the J-bolts directly in the wet concrete. Use 1/2- x 10-inch-long J-bolts. Since you don't have to allow for the thickness of the brick, leave approximately 5 inches of the bolt protruding from the top of the concrete. Be aware that code requirements for concrete slabs vary. Be sure to check that you are in compliance before starting to build. If you are building on the slab, you may be required to provide extended footings along the perimeter, as shown in figure 7.

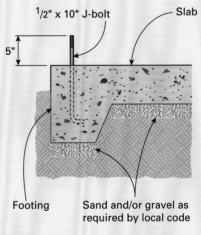

1/2" x 10" J-bolt — Slab

5"

Footing — Sand and/or gravel as required by local code

FIGURE 7

CHAPTER FIVE

Building
the Floor

Congratulations! The tough, heavy, dirty part of the job is done and you're ready to start my favorite part—framing the gazebo. One reason building a gazebo is so much fun is that there is no standard way to do it. Gazebos are a chance to get creative as a builder. There are, however, common, tried-and-true approaches to assembling the basic structure. In this chapter I'll take you through the pros and cons of the most common and practical strategies for installing the posts, framing the floor, and installing flooring. Along the way, I'll introduce you to some innovative twists devised by myself and other builders, including a few we came up with in designing the gazebo plan you'll find at the back of the book (see page 146.) To wrap up, I'll show you how to calculate and build a set of stairs, should your gazebo need one.

the doubled-beam plan

The doubled-beam plan is a very straightforward way to build the floor for a four- or eight-sided gazebo. It's not practical for six-sided gazebos because you can't run two sets of doubled beams parallel to each other with six posts. Doubled beams work well with posts set in holes or on piers. In either case you let the posts *run wild*, meaning that you extend them beyond their ultimate height. Keep in mind that if you are build-ing the gazebo on a sloping site, some of the posts may need to be longer than others. To determine the position for the bottom of the beams, you'll mark level lines on the posts. After you build the floor, you'll cut all the posts off level to each other. Let's take a step-by-step look at this system. In this sequence, I'll talk about building a gazebo with eight sides. For a four-sided variation of this plan, see Framing for a Four-Sided Gazebo, page 59.

Erecting the Floor and Posts

Some gazebos have beams that support light joists. Others forgo beams in favor of heavier joists since the spans on most gazebos are not long. In choosing whether to use beams, it's important to think about what kind of flooring you will use. For example, ¾-inch-thick tongue-and-groove mahogany porch flooring should span no more than 16 inches between supports. You can see right away that if you are going to need lots of attachment points, it makes more sense to have a few beams with many lightweight joists than to have lots of heavy joists.

The next consideration is whether you will run your flooring straight across the gazebo, or use a concentric pattern with boards get-ting longer as you work toward the perimeter. A series of parallel joists that cross support beams makes sense for floor boards that run straight across the joists. But con-centric flooring calls for concentric framing, which is easier and more economical to do using joists alone.

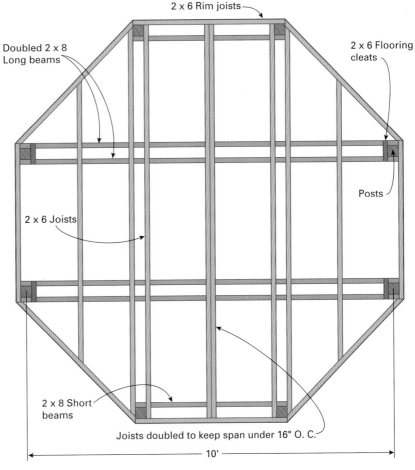

FIGURE 1
Doubled-Beam for Eight Sides

1 **Plumb or install the posts.**
If you are using posts in the ground, put enough dirt in the hole to make them stand up plumb. Don't pack the dirt in too tightly or it will prevent you from adjusting the posts' positions a bit later on. If your posts will be attached to adjustable post bases, first put the post bases in place. Next use a long piece of 2x stock to align opposite bases before tightening the nut. Then nail the posts into the bases.

2 **Determine the floor height.**
The bottom of the beams should be at least 3 inches off the ground—if you use post bases, they'll cover the first three inches of the post. To determine the height of your floor, add the three inches to the width of the beams, the width of the joists, and the thickness of the flooring. For example, 2 x 8 beams are 7¼ inches

wide, 2 x 6 joists are 5½ inches wide, and 5/4 tongue-and-groove flooring is 1 inch thick. Adding 3 inches to these measurements determines that your floor will be 16¾ inches above the highest point of the ground's slope.

3 **Strike level lines.** If your site slopes, find the post that sits on highest ground and mark it where you want the bottom of the beam to be. This is your benchmark. Now mark this same level on all other posts. There are a few ways to do this. You can go high-tech and rent or borrow a laser level. Or you can use a water level, which consists of two clear plastic tubes screwed onto each end of a garden hose. To use it, strap one tube to the post with the benchmark, then fill the tube with water until the water level reaches to the benchmark. Since the height of the water in the second tube will be the same as the first, you can mark the benchmark level on the other posts by holding the second tube up to each of them.

If you only need to span 8 to 12 feet between posts, you can use a less accurate lower-tech method that works well for shorter spans.

Make the post-base nuts snug enough so that you can move the bases with a slight tap of your hammer. Lay a beam or joist across the bases as a guide for aligning opposing bases.

Begin by driving a nail halfway into the post at the benchmark. Select your straightest piece of 2x stock that is long enough to span between posts. Rest one end of the 2x on the nail in the benchmark. Use a 4-foot level to level the 2x, then strike a line on the opposing post.

4 **Attach the beams to the posts.**
With the doubled-beam plan, you can use 2 x 8 beams for gazebos that are up to 14 feet wide. As an example, the doubled-beam plan in figure 1 is for a 10-foot-wide gazebo. As the drawing shows, there are four long beams and two short ones. The short beams extend 1½ inches past the posts on both sides to add support to joists. Once you level the beams across the posts, make sure the posts are plumb, then tack the beams into place on each post by using one or two 10d nails. Predrill holes at the four locations where the long beams sandwich the posts. Use two ⅜- x 8-inch carriage bolts to through-bolt the beams to the posts. Use 6-inch bolts to secure the short beams. If your posts are in the ground, now is the time to pack the earth tightly into the hole.

Use a board with a level on top, supported by a nail driven partway into the benchmark, to make a level mark on the next post.

5 **Design the joist spacing.** The number of joists you need depends on the size of the gazebo and the thickness of the flooring. For the gazebo in figure 1, I assumed we'd use ¾-inch mahogany tongue-and-groove porch flooring, which should span no more than 16 inches between the centers of the joists. I decided to place a joist against each side of the four opposing posts because I knew tying into the posts this way would help stiffen the whole structure. When I measured the distance between the centers of the two innermost of the four joists, I found the distance to be approximately 34 inches. When I divided 34 in half I determined the span to be 17 inches, which is pretty close to the 16-inch- on-center requirement. However, figuring a little extra support through the middle of the floor couldn't hurt, I decided to double the center joist to help close up the span. The outermost joists are simply centered between the rim joists and their neighbors on the other side to create a span of less than 16 inches.

Use carriage bolts to attach the doubled beams to the posts. A washer and nut on the other side completes this strong connection.

framing plan for a four-sided gazebo

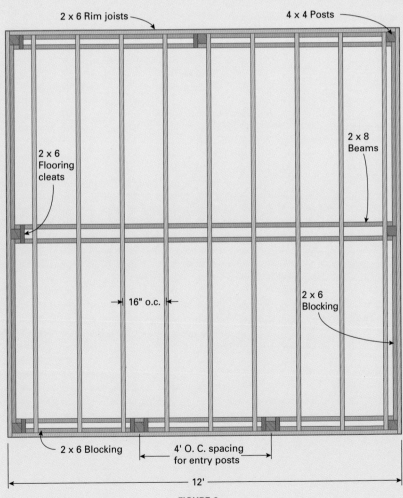

2 x 6 Rim joists

4 x 4 Posts

2 x 8 Beams

2 x 6 Flooring cleats

2 x 6 Blocking

16" o.c.

2 x 6 Blocking

4' O. C. spacing for entry posts

12'

FIGURE 2

Ahhh, the four-sided gazebo. Nary an angled cut in sight. Until you get to the roof framing, that is. But you have a couple of chapters to go before you have to worry about that.

As shown in figure 2, the plan is for a gazebo with four 12-foot-long sides. Its construction sequence is an easier variation of the one for eight sides. There are just two differences to point out:

■ At the front of the gazebo there are four posts instead of the three posts on all other sides. These posts are spaced four feet apart to create a comfortable, attractive entrance.

■ I decided to use blocking to support the joists where they meet the rim joists. Many builders would use joist hangers instead. However, I like blocks better because you can cut opposing pairs to exactly the same length that automatically spaces the joists correctly. The blocks add stiffness, too. All right, I admit it, I simply enjoy cutting nice snug blocks and tapping and nailing them in place, more than I like fussing with metal hardware. (It's a little less expensive, too.)

At each corner, one rim joist is cut square and the other meets it with a 45° cut.

6 **Install the joists.** Note that two of the joists that get nailed to the posts require 45° cuts on both ends. Attach these four joists with four 16d galvanized common nails at each connection. Use 10d nails to toenail them into the beams. Next add the rim joists that have 90° cuts on the ends with two 16d nails into each post. Add the four rim joists that have 45° angles on their ends, attaching them with two 16d nails into the ends of the mating rim joists. One end of these rim joists meets a 45° cut on a long joist. Add two 16d nails at this connection. Put in the doubled middle joist with 16d nails through the rim joists and 10d nails toenailed into the beams. Finally, make the 45° cuts on the ends of the two remaining joists and attach them the same way.

7 **Install Flooring Cleats.** Flooring boards need to be supported wherever they butt into a post. This support is supplied by four 2 x 8s that rest on the doubled beams and are nailed into the posts. Depending on the size of your gazebo and the joist spacing required by the flooring, you may need joists to replace these cleats.

making an angled cutting jig

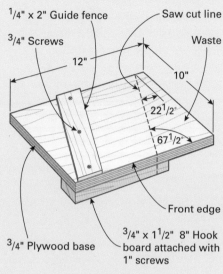

FIGURE 3

After building an eight-sided gazebo, you'll be able to cut 22½° angles in your sleep, and you'll surely be the person selected for equitable pizza slicing. If you're building six sides, then the 60° angle will be your forte.

If you are lucky enough to have a sliding miter saw, you probably won't need this jig, although even then it is sometimes easier to bring a portable circular saw to the work than it is to bring the board to a stationary miter saw. But if you have only a standard miter saw that won't cut through a 2 x 8, or you are doing the whole job with a circular saw, you'll find that the half hour and few scraps this jig requires will be excellent investments.

As shown in the figure 3, the jig has three parts: a $\frac{3}{4}$-inch plywood base, a $\frac{1}{4}$-inch plywood guide board, and a hook board of 1 x 2 ($\frac{3}{4}$- x 1$\frac{1}{2}$-inch) stock to hold against the board you are cutting. You cut the jig to its final width after assembly so the cutting line will be automatically aligned to your saw blade. The jig shown here is wide enough to cut 2 x 8s. If you will be using wider boards for your gazebo, make the jig wider.

1 **Lay out the guide-fence angle.** When building an eight-sided gazebo, we talk about a 22$\frac{1}{2}$° angle, which is really an angle that is 22$\frac{1}{2}$° less than 90°. This means that when you use a protractor to draw a line for positioning the guide fence, you want an angle that is 67$\frac{1}{2}$° from the front edge of the plywood base. (For a six-sided gazebo, the angle is 60°.) Draw the line approximately 2$\frac{1}{2}$ inches from the left side of the base.

2 **Attach the guide fence and hook board.** For the guide fence, start with a piece of plywood that's approximately 14 inches long. Attach it with carpenter's glue and clamps or with $\frac{3}{4}$-inch screws, making sure it is positioned along the layout line. Saw the fence flush to the front and back. For the hook board, use a 1 x 2 that's approximately 8 inches long. Make sure it is flush to the front and left side, then glue or screw it to the bottom of the base. (If you use screws, make sure none will be in the path of the blade when you cut the jig with your circular saw.)

3 **Make the cutting edge.** Hook the jig over the end of a piece of lumber. Hold the base firmly against the guide fence and cut through the jig as shown in the photo at left. Use this new edge to lay out angled cuts by aligning the cutting edge. When you align the cutting edge to a layout line, you know your saw will cut exactly on that line.

Saw through the jig to define the cutting edge.

Align the cutting edge to the layout line to cut the stock.

concentric joist framing

The largest dimension lumber required for the concentric joist frame is 2 x 8s. If you are building a bandstand gazebo that is wider than 14 feet, you can still use 2 x 8s if you add a pier and short post under the center of the floor. Although the infill framing is labeled as 2 x 6 and 2 x 4 in figure 4, you can use larger dimension lumber if you have 2 x 8 or 2 x 6 offcuts you want to use up.

The framing described in the steps below, and shown in figure 4 is for a 12-foot-wide gazebo that uses 5/4 mahogany tongue-and-groove flooring. This flooring can span up to two feet between supports, as shown in figure 5 on page 64.

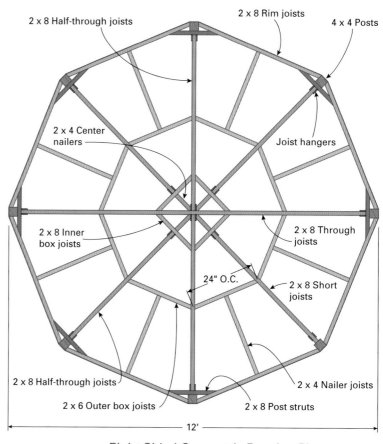

FIGURE 4. Eight-Sided Concentric Framing Plan

Labels in figure: 2 x 8 Half-through joists; 2 x 8 Rim joists; 4 x 4 Posts; 2 x 4 Center nailers; Joist hangers; 2 x 8 Inner box joists; 2 x 8 Through joists; 24" O.C.; 2 x 8 Short joists; 2 x 8 Half-through joists; 2 x 4 Nailer joists; 2 x 6 Outer box joists; 2 x 8 Post struts; 12'

Each rim joist meets a post at 22½°. Make each connection with two 16d nails.

1 **Plumb or install posts and strike level lines for the joist bottoms.** This is done in exactly the same way as described in the first three steps of The Doubled-Beam Plan (see page 57). The only difference is that the level lines you strike on the posts will be for the bottom of the joists, since there are no beams in this plan.

2 **Install the rim joists.** Cut your rim joists with 22½° cuts on both ends. Size them so that their outside faces meet the outside corners of the posts. Secure them with two 16d nails in each post. Your posts should now be plumb and in the right positions. Check that they are, then pack dirt in the post holes or tighten the nuts on the post bases.

3 **Install the 2 x 8 post struts.** You could probably get away without the post struts by nailing joist hangers directly to the posts. I like to use post struts because the joist-hanger flanges just barely fit across a 4 x 4 post, and it makes me nervous to nail these crucial connections so close to the post edges. An easy way

to find the length of the post struts is to set a piece of stock across the bottom of two rim joists and against the inside of a post. Then scribe lines where the strut crosses the joists. Secure the struts with three 16d nails into the posts and two 3-inch screws through each end into the rim joists.

Place a post strut under the rim joists and against a post to scribe the cuts on both ends.

Use joist hangers to attach the half-through joists to the through joists.

4 **Install the through joist and the half-through joists.** Because you know all the post struts are at the same level, you can go ahead and install joist hangers on each one. First install the through joist that runs across the gazebo. Next install opposing joist hangers centered on both sides of the through joist. Then install the two half-through joists between these hangers and the hangers on the post struts.

5 **Install the inner box joists.** The exact length of the inner box joists isn't important as long as they are all the same. I chose to make them 22 inches long, a convenient length that gave me plenty of room to attach joist hangers to their faces. Cut 22½° angles on both ends of the inner box joists and secure them with three 10d nails at each connection.

6 **Install the short joists.** Install joist hangers centered on the inner box joists. Install the short joists between these hangers and the hangers on the post struts.

7 **Install the outer box joists.** Like the inner box joists, the 2 x 6 outer box joists get 22½° angles cut on each end. They are located at the point where the span between joists becomes 24 inches on center, since this is the maximum span of the flooring. If everything is square, cutting them to 22½ inches long should put them in exactly the right spot. Install them with three 3-inch screws at each connection.

8 **Install the nailer joists.** The nailer joists provide midspan support for floor boards in the outer areas where they would otherwise span more than 2 feet. Because the nailer joists span less than 3 feet, they need only be 2 x 4s. Center them between outer box joists and rim joists and secure them with two 3½-inch screws or 16d nails through both ends.

9 **Install the center nailers.** The center 2 x 4s provide nailing for a center medallion or the small pieces that will make up the center of the concentric flooring pattern. On each board, cut both faces of one end at 45° to form a point. Cut the other end square for a snug fit. Put them in place and secure with a few toenailed nails. With a snug fit, you don't need much attachment since these pieces carry very little load.

The center nailers provide nailing for the small center flooring pieces. They are cut to a point at one end to meet the intersection of the through joists and half-through joists.

The inner box joists meet the half-through joists at 22½°. The short joists will meet the inner box joists at 90°.

concentric framing for six sides

Figure 5 shows a framing plan for a 14-foot-wide six-sided gazebo. This plan limits flooring spans to 2 feet for either straight- run or concentric flooring.

The six-sided plan does not have the inner box joists found in the eight-sided plan. Instead, all the joists converge in the center where they are supported by a short 6 x 6 post. All the posts in this gazebo are 6 x 6s.

Like the eight-sided version, two half-through joists meet the through joist at 90°. Attach one of these by nailing through the side of the through joist. Toenail the other into place and then toenail all three joists to the post. Four additional half-through joists come to the center at a 45° angle to the through joist and the 90° half-through joists. These four joists get 45° cuts on both sides of one end to meet the

center intersection. Toenail these to the 90° half-through joists and to the post.

Notice that the 90° half-through joists meet the rim joists, and not the posts. This means you need to install the rim joists right after you get the through joist in place, instead of waiting as in the eight-sided plan. Also, rim joist blocks are added to stiffen the rim joists and strengthen the joist to rim connections.

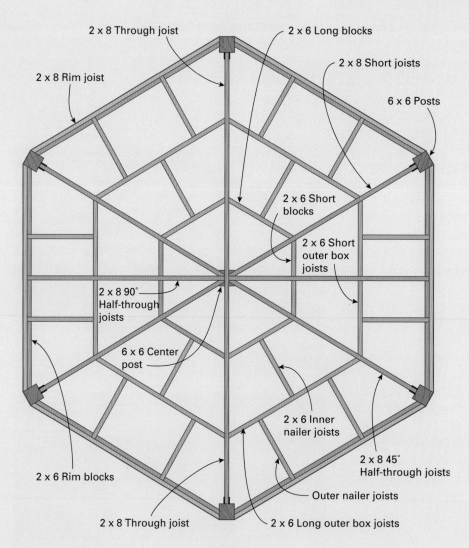

FIGURE 5

The joists are cantilevered 18 in. over the short posts to hide the piers.

cantilevering the concentric joist plan

Most gazebo plans call for running the posts through the floor to support the railings and roof. For the gazebo we built for this book, we decided it would be nice to hide the piers by tucking them 18 inches back from the rim joists. To do this, we supported the floor on short posts placed directly under the through joists, half-through joists, and short joists. At the end of these eight joists we installed a post to support the railings and roof. As you'll see, the posts are built from three pieces that will lap over the sides and front of the joists.

This arrangement eliminated the need for post struts. Other than that, the frame was identical to the concentric joist plan with through posts. However, our construction sequence was a little different. We couldn't install the rim joists right away because the long posts had to be slipped over the ends of the joists first. This was no problem since we didn't have to worry about getting through posts exactly plumb and in the right spot. Plus, leaving the rim joists for later meant we didn't have to keep stepping over them as we worked.

1 **Install the short posts.** Because our gazebo was on a sloped site, we started by installing the short post on the highest pier. This post was only three inches tall; just enough to clear the sides of the post base. A post cap secured the joist to the post. This is the time to attach the post cap to the post.

When hammering nails to attach a small block to the base, swing two hammers to strike the opposing side of the block at the same time as you strike the nail.

A piece of lumber placed from a through joist to half-through joist gives you the height for the top of the short joists.

2 **Determine the height of the opposing short post.** Select your straightest 12-foot-long piece of lumber and place one end on the shortest post. Have a helper hold the other end over the opposing post base. Place a 4-foot level on the lumber. When the lumber is level, measure the distance from the bottom of the lumber to the post base.

3 **Install the through joist.** Nail post-cap connectors to the posts that will support the through joist. Because the joists for our gazebo continued past the posts, we chose to let them run wild at both ends. Later, we cut them all to length at the same time, measuring from the exact center of the frame. This eliminated any inaccuracy that may have happened when we placed the inner box joists and the hangers for the half-through joists. This choice was a matter of personal preference. If you feel more comfortable making cuts on a sawhorse, you can cut the joists to length before you install them. In either case, center the through joist over the posts and nail them to the post-cap connectors.

4 **Install half-through joists and their posts.** Attach two joist hangers, one on each side, both centered on the through joist. Place one end of each half-through joist in its hanger, then level the joist over its pier. Measure for the length of the post, cut the posts to that length, and install them. Add post-cap connectors and install the half-through joists.

5 **Install inner box joists, short joists, and remaining short posts.** Install the inner box joists and joist hangers, exactly as described in step 4 of Concentric Joist Framing (see page 62). You don't need a level to determine the length of the four remaining support posts—just place each short joist in its hanger and put a straight piece of lumber across the joists on each side. Raise the short joist until it touches the bottom of the lumber, and take your post length measurement. Install the posts, post caps, and joists.

Place one end of a board atop the shortest post. When the board is level, take a measurement for the height of the opposing post.

With the hook of your tape on a centered screw, it's easy to mark the joists for cutting to length in place.

6 **Cut the joists to length and install outer box joists.** At the exact center of the intersection of the through joist and the half-through joists, drive a screw partway in. Hook your tape on the screw and walk around the frame marking the final length of the eight joists. Cut the joists to length with a circular saw. Install the outer box joists, as described in step 6 of Concentric Joist Framing (see page 62).

7 **Construct the posts.** The posts that support the railings and roof were each made of a 2 x 4 sandwiched between two pieces of 5-inch-wide 5/4 lumber. The 2 x 4 was centered between the side pieces, creating ½-inch reveals that add visual interest. As shown in figure 6, the 2 x 4 was notched to cover the front of the joist. It was also cut to the height of the ceiling, which consists of 2 x 4s that were later sandwiched between the 5/4 side pieces along with the rafters. For now, let the 10-foot-long side pieces run wild. We assembled the posts with pairs of 2-inch screws set every foot—we offset the screws because we liked the way it looked.

It's easy to pick the out-facing sides of the post boards when you lay them on the deck. For a consistent screw pattern, use a straightedge to draw a light line across the boards every foot.

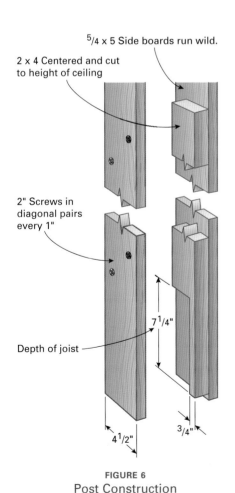

⁵/₄ x 5 Side boards run wild.

2 x 4 Centered and cut to height of ceiling

2" Screws in diagonal pairs every 1"

7¹/₄"

Depth of joist

4¹/₂"

³/₄"

FIGURE 6
Post Construction

Slip the preassembled posts over the ends of the joists and attach with five screws on each side.

Install each 1 x 8 skirt board with five pairs of 8d finishing nails.

8 **Attach the posts to the joists.** Slip each post over the end of a joist, check that it is plumb and that the post notch is against the end of the joist. Then attach the post to the joist using five 2-inch screws in each end.

9 **Install the rim joists, nailer joists, and skirt boards.** The rim joists get a 22½° cut on each end. Cut them to length so that they will be recessed 1¼ inches from the front of the posts. Install the rim joists with three 16d nails at each connection. Cut the 2 x 4

nailer joists and install them with two 10d nails at each connection. Now cut and install the skirt boards with 8d galvanized finishing nails—by installing ¾-inch-thick skirt boards, you will carry the ½-inch reveal of the built-up posts through to the bottom.

got a little wobble?

Sometimes when you use post bases your floor frame may be slightly wobbly even after it is completely framed. This is especially likely if the frame is more than a foot or so from any of the piers. If you find this to be the case, two or four cross-braces nailed to the inside of the post and meeting the side of a joist will make things rock solid. Make the braces from 2 x 4s. Install them on the posts where the frame is highest off the piers. For the gazebo built for the book, we used four cross braces on two posts. The wobble stopped after we installed two cross braces

Installing the Floor

For your gazebo floor, you can use softwood lumber decking boards or tongue-and-groove porch flooring. As discussed in Chapter 3 (see page 35) decking can be 2 x 4s, 2 x 6s, or 5/4 x 6 decking, which is approximately 1-inch thick with rounded edges. Tongue-and-groove porch boards are usually mahogany and are available as 1xs or as 5/4 boards. All of these materials can span two feet, except, as I mentioned, the 1x tongue-and-groove boards.

To attach decking boards, use galvanized decking screws that are long enough to penetrate at least 1 inch into the joists. Use two screws wherever a board crosses a joist.

installing a straight-run floor

The strategy for installing a straight-run floor is almost the same whether you are installing decking or tongue-and-groove flooring. Run the boards perpendicular to the joists, and, unless you're building a rectangular gazebo, cut the boards off at an angle. The easiest way to handle this is to let each board run wild by a couple of inches on each end, then after they are all installed you can cut them off at the same time.

Decide how much you want the floor to overhang the skirt board. There's no rule about this. I like the looks of a ¾-inch overhang with 5/4 boards. Perhaps you'll want a little less overhang for 1xs or a little more for 2x decking.

For an eight-sided gazebo, you may have to notch the outermost boards to fit around two posts. You will definitely need to notch the second course of boards. Always install decking board with a little space between them—use 10d nails placed between the boards as spacers. Always install tongue-and-groove boards tightly together. Work your way across the gazebo, notching the ends of boards as necessary to fit around posts.

When all the boards are installed, snap lines across the ends and use your circular saw to cut them all to the desired overhang. If you want to round over the overhang, use a router, a belt sander, or a block plane and a piece of sandpaper. If you use a router, you'll probably want to round over the boards that are notched around posts before you nail them down. If you don't, the posts will get in the way of the router base.

blind nailing tongue-and-groove flooring

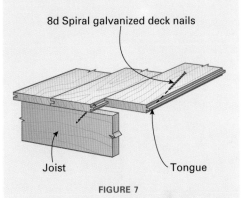

8d Spiral galvanized deck nails

Joist

Tongue

FIGURE 7

When you install tongue-and-groove flooring, rip the groove off the outermost course of boards. This ripped edge becomes your overhang. Clamp the board to the rim joist. Now use 8d spiral galvanized deck nails and blind nail through the tongue at an angle back through the meat of the board and into the joist, as shown in figure 7. Use a nail set to drive the nail below the surface.

Put the next board in position. If you find you need to tap a board to get it fitted snug to the one before, fit the groove of a scrap piece of flooring over the board's tongue and tap on the scrap. This way you won't smash the tongue with your hammer.

notching flooring around built-up posts

To fit the flooring boards around the built-up posts, we started by ripping the groove off the first course of boards, then rounded that edge with a router. We cut the board to a length that would let it overhang the skirt board by approximately 1 in. then nailed it in place. Here, the nail is being set.

Before notching the second course of boards, we cut them to length with a 22½° cut at both ends. We notched them square to their ends to fit around the center 2 x 4, then used a flooring scrap as a guide for sawing the top of the notch.

After sawing the bottom of the notch flush with the top of the joist, it was a simple matter to knock out the soft cedar waste with a chisel.

installing concentric flooring

If you're using decking boards, start at the outside. This way, you notch the first boards to fit around the posts and start with a nice even overhang. From there, work your way around and toward the center, marking and cutting each board to fit as you go.

However, with tongue-and-groove boards, starting from the outside creates a significant complication. Because of the angled cuts on the ends, the boards are longer on the outside edge than the inside edge. This means you can't just make the usual cuts and slide the last board in each course into its tongue. In most cases it is best to start by making a six- or eight-sided center medallion. (see page 72, Making a Center Medallion). Position the medallion very precisely in the center of the floor, making sure its sides are parallel to the skirt boards. When you get to boards that have to be notched for the posts, undercut the back of the notches. This enables you to lever the board around the mating tongue. (See figure 8.)

Despite the fitting problem, there are cases where you may want to start concentric tongue-and-groove boards from the outside. We decided to take this approach for the gazebo built for this book because of its built-up posts. We felt it would have been too complicated to notch the flooring boards around the reveals created by the 5/4- x 5-inch boards that sandwich the 2 x 4 at the center of each

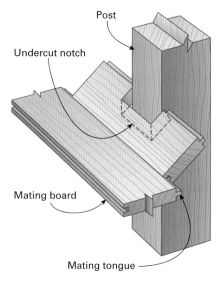

FIGURE 8
Tongue-and-Groove Flooring Around a Post

post. Instead, we notched the 5/4 boards to allow the flooring boards to slip under them far enough to butt into the center 2 x 4. As you can see in the sidebar, Notching Flooring Around Built-Up Posts (see page 70), this precluded starting the floor in the middle.

The surefire way to fit a concentric pattern is the same for decking or tongue-and-groove boards: Cut one end of the piece to the required angle, put the piece in place, and then mark for the cut on the other end.

When you run into a problem fitting flooring boards around built-up posts, the easiest solution is to cut one board in each course in half where it lands on a nailer joist. This really isn't a big deal—if you cut a scarf joint and angle it approximately 25° it will hardly be noticeable. Of course, this solution only works when you have nailer joists to land on, which is no longer the case once you get past the outer box joists. Following is another solution that eliminates the need to cut boards in half.

When marking boards in place for cutting, slant the pencil mark so there will be no question which way to angle the cut when the piece is brought back to the saw.

 ## fitting the last tongue-and-groove board

In a concentric-patterned floor, like the one in the gazebo built for this book, the boards are shorter along the inside edge than they are along the outside edge. As a result, you can't just slide the last board in each course into place in the normal way. You have to lower it into place, as shown in the photo.

To make this possible, we undercut both ends of the last board, as shown in figure 9. The angle of the undercut isn't crucial—approximately 30° works well. This is actually a compound angle cut, since you cut the top of the board at 22½° for an eight-sided gazebo, or 60° for a six-sided gazebo.

To meet the undercuts on the last board, you must overcut the boards that meet it on each side with a mating angle, as shown in the figure 9. Now you'll need to make one more little cut on the two overcut boards. Here's why: When you cut the boards square to their faces, they end over the cen-

When working toward the middle to install a concentric tongue-and-groove floor, the last board in each course must be lowered into place.

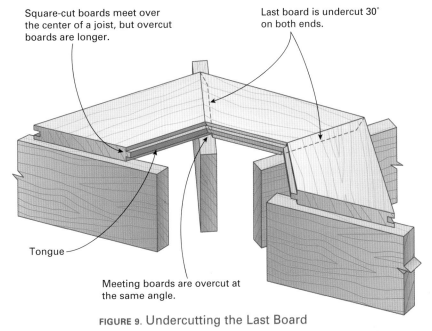

Square-cut boards meet over the center of a joist, but overcut boards are longer.

Last board is undercut 30° on both ends.

Tongue

Meeting boards are overcut at the same angle.

FIGURE 9. Undercutting the Last Board

ter of joists. When you overcut the boards, the top of the cut is over the center of the joist, but the bottom of the cut is a little longer. To solve this, you simply nip back the tongue end of these boards, as shown in the photo at left. Since the last board covers the nip, you'll see how much you need to nip the first time you fit these boards.

Each of the two boards that meet the last board in a course gets overcut on one end. Note how the overcut is nipped back on the groove side.

making a center medallion

There are any number of ways to finish the center of a concentric flooring pattern. The one you choose depends to some extent on the flooring material you're using. However, it mostly depends on what you like.

The easiest approach is to cut triangles that fit the space, then screw them into place as shown in the photo. This idea is probably best reserved for six-sided floors with 5- or 6-inch-wide decking boards. With 2 x 4 decking, or with eight sides, the center pieces may be too small to attach with screws without splitting. (You should certainly predrill in any case.) If you decide to use this approach with tongue-and-groove flooring, you'll need to glue the pieces in place. Polyurethane glue is excellent for this.

If you're using 2 x 4, 2 x 6, or 5/4 decking, you can cut six or eight sides onto an 8- or 10-inch-wide piece of the same thickness. This one-piece medallion avoids the problem of how to attach a bunch of little pieces.

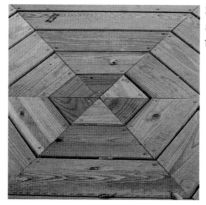

Six triangles fitted together is a straightforward way to bring the floor together at the center.

Another really easy solution is to use a circular center medallion. When you get to the last course, tack the last pieces in place, as shown in figure 10. Then tack a scrap of flooring in the center space. Find the center of the floor, and use a compass to scribe a circle around the last course. Cut the arc on each board. Use the same compass setting to lay out the circular medallion on stock that's the same thickness as the flooring. This could be a single wide board, two pieces of flooring stock laid side by side, or several pieces of tongue-and-groove fitted together.

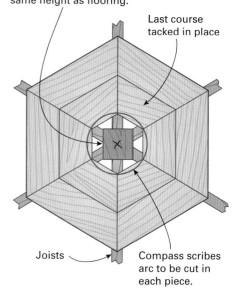

Flooring scrap, tacked in place, puts compass point at same height as flooring.

Last course tacked in place

Joists

Compass scribes arc to be cut in each piece.

FIGURE 10
Scribing Cuts for a Circular Medallion

The turtle tiles are traced onto the center of the gazebo floor.

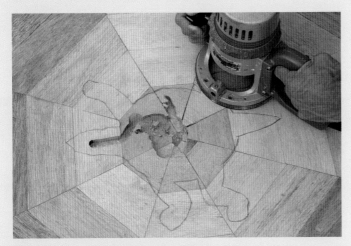

The recess for the turtle is routed out freehand using a straight bit set to the correct depth.

As we constructed the gazebo for this book, a turtle would occasionally pop his head out of the pond to check our progress. Or maybe he was just trying to snag one of the dragonflies that frequent this site. In any case, these visitors inspired Daniel Grant to finish the center of his gazebo in a whimsical way. He visited a local craftsperson, Natalie Surving, who makes ceramic tiles in the shape of animals, and returned with six tiles that form the shape and colors of a turtle, and one shaped and colored in the form of a dragonfly.

After gluing the center tongue-and-groove pieces into place, he sanded off the excess glue. Then he traced the shape of the turtle at the center and the dragonfly prey a few boards away.

Daniel removed the tiles and followed the outlines on the floor with a router equipped with a straight bit. He set the depth of the bit so the tops of the tiles would be just slightly higher than the floorboards, and guided the router by eye. Once the floor was finished, he glued the tiles in place.

A tile center medallion in the shape of a turtle pursues his dragonfly prey across the gazebo floor, reflecting life in the pond a few feet away.

Building Stairs

If the entrance to your gazebo is more than 8 inches from the ground, you'll need to build a set of steps or some sort of stepping platform. In this section, I'll show you how to calculate and build a basic outdoor step and describe one design for a platform.

When designing any steps, the deeper you make the treads, the shorter the risers should be. The depth of the tread (the part you step on) minus the nosing (more on that later) is called the unit run, while the height of the riser (the vertical space that's perpendicular to the tread) is called the unit rise. You don't want a unit rise of more than 8 inches or a tread less than 10 inches deep. For a comfortable stride, the unit run and unit rise should equal approximately 17 or 18 inches.

Jim Quinlan's gazebo steps

designing a run of stairs

Outdoor stairs typically have a smaller unit rise and, consequently, a deeper unit run, than you'll find on indoor staircases. This is due to the fact that when building outdoors you can usually make the bottommost step as far from the gazebo floor as you like. Besides, gradual stairs look more appropriate in a landscape. You should aim for a unit rise of approximately five to seven inches and a run of approximately 11 to 13 inches.

1 Choose a proposed total run. First, decide approximately where you want the stairs to land. Let's say, for example, that the entrance to your gazebo is 17½ inches off the ground. If you want each step up to be about six inches, that means you need three steps. Since you count the gazebo floor as one step, you will need two more steps down. Using a typical unit run of 12 inches, that means you will need a total run of about 24 inches. Actually, at this point we'll call it the *proposed total run*, because you may decide to adjust it later.

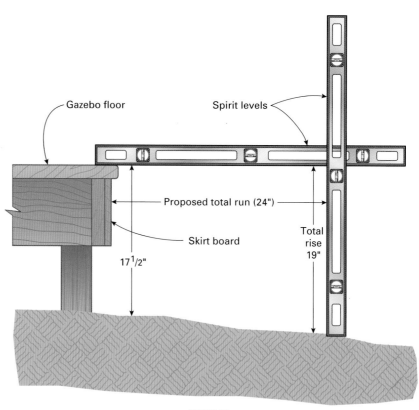

FIGURE 11
Finding the Total Rise

2 Find the total rise. In our example, the vertical distance from the top of the floor to the ground is 17½ inches. But that doesn't mean the total rise is 17½ inches because the ground elevation where the steps end is probably different than the elevation at the edge of the gazebo. To find the true total rise, rest one end of a level on the gazebo floor and extend the other end of the level out past the proposed total run. If the level isn't long enough, place it atop a straight 2 x 4. Measure out the proposed total run along the level and place another level crossing that point and extending to the ground. Note: Measure the total run from the skirt board under the flooring. Like the tread nosings, the floor overhang doesn't count in the calculation. Make sure the level is plumb, and then measure the total rise, as shown in figure 11. If the level isn't long enough, you can use it to plumb a straight 2 x 4.

3 Calculate the unit rise. Let's say, for example, you find that the total rise is 19 inches. Divide 19 by the three risers and you get 6.33. With the help of a decimal conversion table or a construction calculator, convert this to a unit rise of 6 $\frac{5}{16}$ inches. Add 6$\frac{5}{16}$ to a unit run of 12 inches and you get 18$\frac{5}{16}$ inches, a little over the combined maximum of 18 inches, but not enough to worry about. Now your proposed total run of 24 inches can become you final total run.

4 Adjust the proposed total run, if necessary. But what if you found the ground sloped a little more, and that, for example, your total rise turned out to be 22½ inches? In this case, 22½ divided by three gives us a unit rise of 7½ inches. That's still a comfortable step up, but not when combined with a 12-inch unit run. A unit run of 10½ inches added to a 7½-inch unit rise would give us a comfortable total of 18 inches. In this scenario, our final total run is 21 inches.

5 Add a step, if necessary. Sometimes you'll find you need to add a step or even two to the total run, especially if the ground slopes a lot. Let's say in our example the ground slopes so that our total rise turns out to be 28 inches. Dividing that by three results in a rise of 9⅝ inches—no good since it's way beyond a comfortable rise. So let's divide 28 by four steps to get a unit rise of 7 inches. That's a good riser height if we use an 11-inch unit run. Discounting the floor itself, we now have three unit runs of 11 inches for a total run of 33 inches.

dealing with sideways slope

One of my first jobs as a freelance carpenter was building a new front stoop for a couple of friends. I'd read up on how to do it just as you are reading now. And I calculated the total rise just as I've told you to do. What I failed to consider was that the sidewalk where the stoop was to land ran downhill at a pretty steep slope. Without thinking ahead, I measured the total rise from the uphill side. As a result, the downhill side of the first step was much too high. I should have measured from the downhill side, as shown in figure 12, and allowed the first step to get shorter as it ran uphill.

Ideally, you should adjust the grade or build a slab so the ground in front of your steps is level. But like my little stoop project, this is not always possible. My friends forgave my error, but I have always felt bad about that stoop I built more than two decades ago. At least now I have the consolation of knowing you won't make the same mistake.

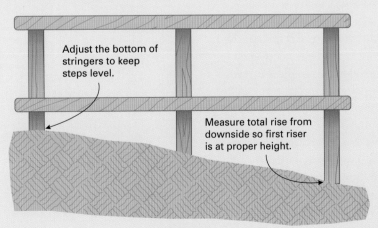

Adjust the bottom of stringers to keep steps level.

Measure total rise from downside so first riser is at proper height.

FIGURE 12

stair design option: adjusting the unit rise

You can always achieve a comfortable step by adjusting the total run. But there are times you may want to put the chicken before the egg by adjusting the unit rise instead. There are two related reasons you might want to do this. First, even if the calculations work, you may feel that a more gradual step would look more graceful with your gazebo. Second, you might want to design your stair treads around standard lumber dimensions.

Stair treads should have a nosing of ¾ to 1¼ inches. The main purpose of the nosing is to prevent you from stubbing your toes on the riser board. But a nosing is a good idea even if you plan to leave the riser spaces open—it makes the steps seem more natural and safe.

Since the nosing overhangs the tread below, it doesn't count in the unit run or total run—that's why you discounted the flooring overhang when you calculated the total run. Of course, the nosing adds to the overall width of the tread, making your tread about ¾ to 1¼ inches wider than the unit run.

I like to make outdoor stair treads from two pieces of lumber, leaving a ½-inch space between each piece. Two narrow pieces are less likely to warp than one wide piece, while the space between provides drainage for rainwater. Since 2 x 6s are actually 5½ inches wide, two of these boards with a ½-inch space would give you a comfortable 11½-inch tread. This would work well with a unit run of about 10½ inches. For an even gentler step, make each tread from a single 2 x 6 and a 2 x 8. With the ½-inch space, this nets you a 13¼-inch tread— perfect for a unit run of 12 inches with a unit rise of 5 or 6 inches.

choosing a stringer design

The boards that support the treads (and risers, if you will be using them), are called stringers. Usually you'll use two stringers—one on each side of the stairs. But if your stairs are more than 3 feet wide or 5 feet tall, you'll want to stiffen them up by adding a center stringer.

You can use *open stringers* or *closed stringers*, as shown in figure 13. With open stringers, the boards are cut away to accept the treads and/or risers. (Center stringers are always open.) With closed stringers, the treads and/or risers are attached to wood or metal cleats fastened to the inside faces of the stringers. Alternately, closed stringers can be grooved with dadoes to accept the treads and/or risers.

When building outdoor structures, you want to minimize joints that can trap water, since collected water promotes rot. To prevent this, it's best to use open stringers, especially without riser boards. If you prefer the more refined look of closed stringers, I recommend using metal hardware instead of wood cleats for attaching the treads—the horizontal joint between the bottom of the treads and the top of the cleats is particularly prone to rot. You'll find that the metal hardware is easy to use and will be hidden. I'd avoid using dadoes unless the stair will be out of the weather. While dado connections are the most elegant of all, they are the most vulnerable to rot.

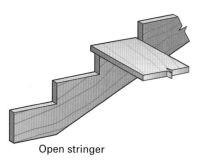
Open stringer

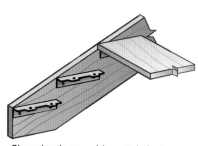
Closed stringer with metal cleats

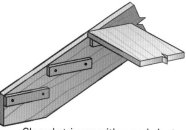
Closed stringer with wood cleats

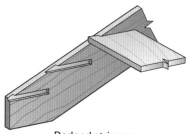
Dadoed stringer

FIGURE 13. Stringer Options

Regardless of the type of stringers you use, you lay them out the same way using a framing square. You should use nothing narrower than 2 x 12 boards to make open stringers, since this design removes a lot of the structural meat from the boards. You'll need at least 2 x 10s for most closed stringer designs, just so you have enough width to attach the cleats. To get a rough idea of how long your stringer boards need to be, run your tape measure from the gazebo floor down at an angle to the point you marked for the total run, then add about six inches to that measurement.

1 **Make the first mark.** Lay the stringer board on your sawhorses and make the first mark approximately 6 inches from one end, as indicated in figure 14. If you make the mark closer to the end of the board, you may not have enough wood to make the bottom riser.

2 **Lay out the unit runs and unit rises.** The longer leg of a framing square is called the blade; the shorter leg is the tongue. As shown in figure 14, align the unit-run measurement on the outside of the framing square's blade to the first mark, and align the unit-rise meas-

urement on the outside of the square's tongue to the edge of the board. Draw cut lines for the first rise and first run. Now align the unit-run measurement on the blade to the top of the first unit rise. Again, align the unit rise on the tongue to the edge of the board. Draw in the second unit rise and run. Continue this process for as many steps as you will lay out.

3 **Lay out plumb and level cuts.** To lay out the bottom plumb cut, draw a line extending from the first mark, square to the first unit run. From the first mark, measure down a distance equal to the unit rise minus the thickness of your stair tread stock. From this point, draw the bottom level line that is cut square to the bottom plumb cut. Now, lay out the top plumb cut perpendicular to the top unit run.

4 **Cut the stringers.** If you are making closed stringers, you only need to make the top and bottom plumb cuts and the bottom level cut. Use a circular saw to cut to the rise and run intersections. You'll need to complete the cuts with a handsaw to avoid over-cutting the intersections.

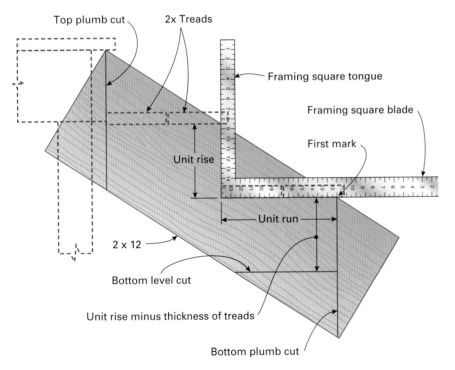

FIGURE 14
Laying Out a Stringer

assembling and attaching the stairs

For open stringers, assembling the steps is simply a matter of screwing the treads to the stringers. To attach the stair to the gazebo, use 3-inch screws to screw the top tread down into a horizontal 2 x 4 ledger attached through the skirt board, if there is one, and into the rim joist. Also use 3-inch screws through the outside of the stringers and into the ends of the ledger.

Often an open stringer will meet a 2 x 8 rim joist at a point that is too far down for adequate attachment. In this case, you need to attach stock across nearby posts to pack out the rim joist, as shown in figure 15. The pack-out consists of attaching a 2 x 4 to two posts using carriage bolts. If there is a 1 x 8 skirt attached to the rim joist, countersink the round bolt heads, then screw a 1 x 4 across the face of the 2 x 4. Use carriage bolts to attach a 2 x 6 ledger so that it overlaps the skirt and the pack-out.

To assemble closed stringers, align the metal stair cleats to the tread lay out lines and screw them in place. Screw up through the cleats into the treads.

The top edge of closed stringers will meet the bottom of the gazebo-floor nosing. This means you can attach the stringer simply by toe-nailing 16d nails through both sides of the stringer, through the skirt board, and into the rim joist. Using four nails in each stringer— two on each side—will give you plenty of strength. If you don't want to look at those nails, you'll get an even stronger connection by using the same attachment method described for open stringers.

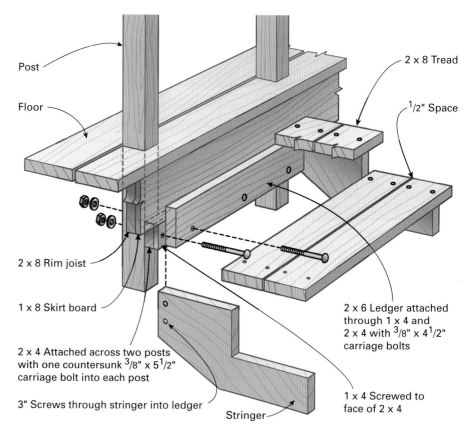

Post

Floor

2 x 8 Tread

1/2" Space

2 x 8 Rim joist

1 x 8 Skirt board

2 x 4 Attached across two posts with one countersunk $3/8$" x $5^1/2$" carriage bolt into each post

3" Screws through stringer into ledger

Stringer

2 x 6 Ledger attached through 1 x 4 and 2 x 4 with $3/8$" x $4^1/2$" carriage bolts

1 x 4 Screwed to face of 2 x 4

FIGURE 15
Attaching an Open Stringer

Even if your stringers are made of pressure-treated or other rot-resistant wood, you should make sure they don't sit on wet ground. You can simply land them on a few inches of gravel, or build a landing of brick, concrete, or stone.

Another option, especially if you are planning to include a handrail, is to bolt the bottom of the stringers to posts on both sides. To do this, begin by anchoring the posts to the ground in the same way as the gazebo posts. Leave the stair posts long until you install the stringers. It's also a good idea to make the first rise of your stringer approximately ½ inch short to raise it out of the dirt. Note that if you use posts attached to the outside of the stringers you'll need to align the outside of the stringers to the inside of the gazebo railing posts, and not to the outside of the posts. If your treads go past the outside face of the stringers, you'll need to notch the bottom tread around the post.

Secure the posts to the stringers with pairs of ⅜- x 5-inch carriage bolts. To lay out the cut for the top of the stair posts, start by marking the top and bottom points on the corresponding gazebo posts, as shown in figure 16. Mark equidistant points on the stair posts. Run a chalk line from the mark for the top of the gazebo railing, down to the mark for the top of the stair post and across the post. Cut the post off along the line. Now that you know this angle, you can use it to adapt any of the railing designs shown in Chapter 7 to turn them into stair rails.

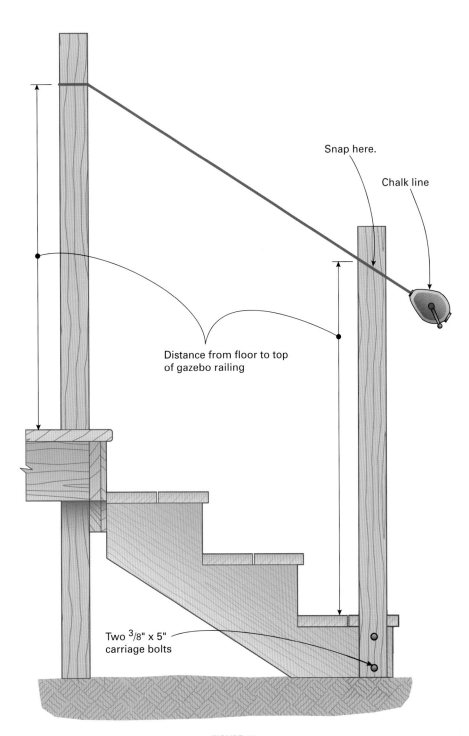

Snap here.

Chalk line

Distance from floor to top of gazebo railing

Two ³/₈" x 5" carriage bolts

FIGURE 16
Finding the Railing Cuts

building an entrance platform

The entrance to the gazebo we built for this book faces a pond. Since no one approaches the gazebo from this direction, we originally assumed we would place the entrance in the direction of typical approach. Because the site sloped down toward the pond, putting the entrance on a different side could have eliminated the need for a step up. But then someone said, "Wouldn't it be nice if the view of the pond was unobstructed by a railing?"

Owner/builder Daniel Grant agreed. Entering from the pond side would require two steps up. It also meant people would walk up to the gazebo and then have to walk between the gazebo and pond before turning and stepping up to go inside. Daniel's solution

The platform cantilevers over two short posts on piers.

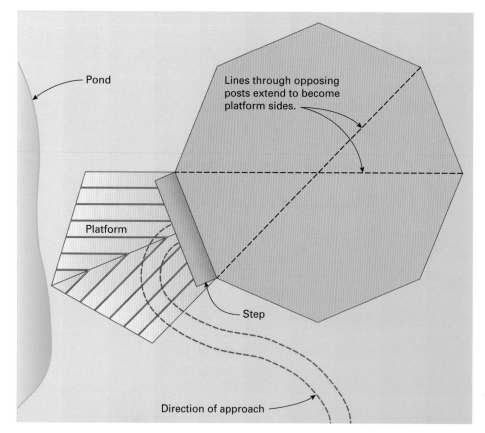

FIGURE 17
Platform Plan

was to provide an angled platform—a small deck, really—with
one step built on the platform.
The platform provided a comfortable way to turn into the step. It
also happened to be the perfect
place for a young boy like Daniel's
son Adam to sit and cast his fishing pole.

An entrance platform like this
is very site-specific. Since there is
little chance you will build one just
like this, I won't bore you with a
blow-by-blow of how it was built.
But seeing how the platform was
designed might inform your thinking. As you can see in figure 17, the
platform got its shape by extending
lines through two opposing posts.
You can see from the photos that
it is a simple frame cantilevered
over two short posts on piers. On
the gazebo side, the platform is
hung from two 2 x 8s nailed to the
inside of the rim joist. Decking
boards meet in the middle with a
V-pattern that runs parallel to the
diverging sides.

The platform hangs on 2 x 8s attached to the inside of the rim joists.

The 5/4- x 4½-in. cedar decking boards run parallel to the sides that diverge
from the gazebo. The boards meet in the middle and are cut off to a snapped
chalk line after installation.

Adam and Daniel Grant
enjoy fishing together from
the entrance platform.

CHAPTER SIX

Building Beams, Plates, Brackets, and Friezes

Once the floor is completed, I like to work on the parts of the gazebo that are below the rafters but above the railings. This locks the posts in a plumb position, so you can build the railings for a perfect fit. It's also easier to move your stepladder around without having railings in the way.

This gazebo, designed and built by Jim Quinlan, uses a header beam. Railing, brackets, and rafters are attached directly to the posts.

FACING PAGE: This gazebo in the town park of Warwick, New York, has boards fastened to the sides of the railings. These boards are, in turn, screwed into the posts. Similar side boards fasten the friezes, brackets, and rafters to the posts.

Although I won't describe how to build railings until the next chapter, you should think about them now. One reason, obviously, is that the railings need to work aesthetically with your decisions about how you will tie the posts and rafters together, and whether or not you will have friezes and/or brackets. Also, from a building standpoint, you should think about how all these elements work together as a system.

Aesthetically, the design possibilities for the walls of your gazebo are virtually endless. The great news is that once you are familiar with just a few construction methods, you can adapt one of these systems to pretty much any design you like. My goal in this chapter is to introduce you to these systems and show you some variations. Once you understand the systems, you'll be able to adapt them to your own gazebo design.

Choosing a System

Let's look at the pros and cons of three approaches—top plates, header beams, and side boards. I'll help you decide whether to include ceiling joists, and then I'll show you how each of the three methods are done.

Jim Quinlan purchased ornate posts for his gazebo and then notched the front of the header beam into the posts for a super-strong connection that eliminated the need for ceiling joists.

using top plates

Start by thinking about how you will attach the railings and rafters. If, for example, you are planning a gazebo that is only eight or ten feet across and will have only hip rafters that land directly over posts, you can use doubled 2 x 4 plates. Nail one layer of top plates to the top of the posts, install the rafters, then butt the second layer of top plates into the sides of the rafters. Attach any friezes or brackets directly to the plates and posts, and the railings directly to the sides of the posts to complete this clean, simple look. Yes, I have seen gazebos with single top plates, but by using only a few more pieces of 2 x 4 why not stiffen the structure and lock in the ends of the rafters?

using header beams

If your plan calls for jack rafters that will land on the plates, doubled 2 x 4s won't provide enough support. You'll need the strength of header beams to support these intermediary rafters. The header beams shown in the photo left were made from two 2 x 6s on edge with a 1 x 6 below. In this system, you notch the posts to receive the headers. This involves more work than the other systems, but you will be rewarded with a rock-solid structure. The notch does an excellent job of shouldering the weight of the roof as well as resisting outward thrust and sideways racking.

ABOVE: This view of the interior of the Warwick Town Park gazebo shows how the ¾- x 3-in. side boards continue up past the top of the posts to capture the rafters. Instead of a bird's-mouth in the rafters, the post tops are cut at the pitch of the roof.

BELOW: A closer look shows how side boards attach railings to posts. Decorative angled cuts on the railing side boards and upper side boards add design flair.

using side boards

If you use plates or beams, you'll need to make bird's-mouth cuts on all the rafters. A bird's-mouth is a notch in the bottom of the rafter that fits around the beam or plate. As you'll discover in Chapter 8 Building the Roof, bird's-mouths are easy enough to cut, but they do need to be accurate.

A third, more forgiving system, shown in the photo above avoids bird's-mouths altogether. Here, 1-inch-thick boards are attached to the sides of the posts. These boards run past the tops of the posts. They provide a way to attach the sides of the rafters, made of 4 x 4s to match the posts, to the posts. In the gazebo shown here, the side boards also provide convenient attachment for brackets and friezes. By screwing the side boards into the ends of the friezes and brackets, then screwing the boards into the posts, you eliminate the need to toescrew or toenail close to the edges of the friezes or brackets. This greatly reduces the risk of splitting.

As shown in the photo below, this same side-board system is used to fasten the railings to the posts. As an added plus, the railings can easily be removed as units for painting, repair, or replacement.

Installing Ceiling Joists

In deciding on a system for tying together rafters and posts, consider whether you will include ceiling joists. The purpose of ceiling joists is to help resist the outward pressure of the rafters on the walls of a structure. For example, ceiling joists in a typical gabled roof, where many long rafters push out in two directions, prevent the walls of the house from being pushed out.

Most gazebo designs, however, don't need ceiling joists, since a gazebo's rafters are shorter and exert less outward pressure. Furthermore, a gazebo distributes the pressure from rafters to at least four sides, or more typically six or eight sides. The sides, in turn, distribute the load to the posts, which are knitted together by the floor framing, railings, and top plates or header beams, as well as any brackets or friezes you might add.

Header beams are better than top plates for resisting outward thrust. Built-in benches can also help. If the posts run through the floor to rest on piers, they will offer more resistance to outward thrust than posts attached to the tops of beams. A surefire way to eliminate the need for ceiling joists is to add extra leverage to your posts by sinking them 3 or 4 feet into the ground.

Still, you might want to include ceiling joists in your design. Perhaps you just like the way they look. Or maybe your gazebo won't have a railing, friezes, or brackets or posts running through to the piers or buried in the ground to help resist rafter thrust. Another more unusual reason is featured in the gazebo

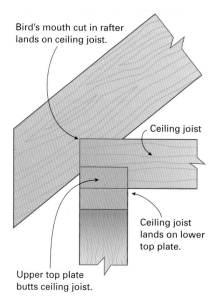

Bird's mouth cut in rafter lands on ceiling joist.

Ceiling joist

Ceiling joist lands on lower top plate.

Upper top plate butts ceiling joist.

FIGURE 1. Ceiling Joist with Doubled Top Plate

plan at the back of this book, where the ceiling joists become floor joists because they support a small platform for a lookout tower. In any case, I can't imagine a situation in which you would need to use anything beefier than 2 x 4s for your gazebo's ceiling joists.

When installing ceiling joists with doubled top plates, first install the bottom plate, followed by the joist, then butt the second plate into the sides of the joists, as shown in figure 1. The bird's-mouth notch in the rafters will land on top of the joists.

In the following steps, I will describe how to install ceiling joists for an eight-sided gazebo, then I'll show you the slight changes necessary for six sides. You don't usually use ceiling joists with four-sided gazebos, though you could have diagonal joists running between the corners. Personally, I think that would look awkward. In that case, I believe it would be better to rely on other methods of resisting outward rafter thrust.

1 **Cut the lapped through joists.** In this design there are two joists that span from one post to another. These joists meet at a 90° lap joint. To get the length of these joists, measure from the outside corners of opposing posts. The lap joint consists of a 1½-inch-wide by 1¾-inch-deep notch cut into the exact center of each joist. You can easily make the notches by setting your circular saw to cut 1¾ inches deep. Make two cuts to define the sides of the notches. Then make several passes between the first two cuts to remove approximately half of the waste. Use a chisel to knock off the remaining waste. See photos below.

Two circular saw cuts define the sides and depth of the half-lap.

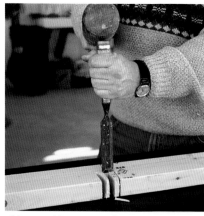

After making several more passes to remove some of the waste, knock out the rest of the waste with a chisel.

2 **Install the through joists.** With its notch facing up, toenail one of the through joists to the top of the post or top plate. Assemble the lap joint and toenail the other joist to the posts.

3 **Cut and install the half-through joists.** Four half-through joists meet in the corners formed by the through joists, as shown in figure 2. To fit in the corners, cut a 45° angle on both ends of one side of each half-through joist. Cut the half-through joists to length and toenail them into the through joists and to the posts.

A half-through joist with the angles cut on the ends fits into the corner of the through joist.

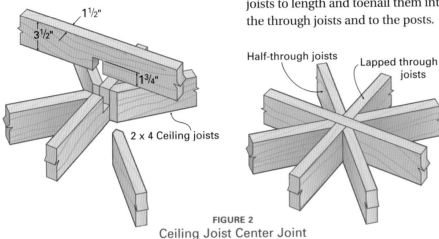

1½"

3½"

1¾"

2 x 4 Ceiling joists

Half-through joists

Lapped through joists

FIGURE 2
Ceiling Joist Center Joint

installing ceiling joists in a six-sided gazebo

For six-sided gazebos, the half-laps are cut at a 60° angle.

As for eight sides, the ceiling joists for a six-sided gazebo start with two joists that span the gazebo and cross with a half-lap joint. The difference is these through joists cross at a 60° angle. This means you need to cut the sides of both half-lap joints at 60° as shown in figure 3. It also means that only two half-through joists meet the corners created by the intersection of the through joists. These get 60° cuts on both sides of one end instead of 45°.

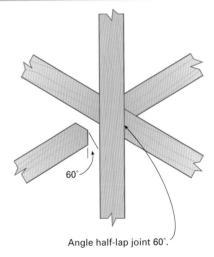

60°

Angle half-lap joint 60°.

FIGURE 3

secret support for a massive roof

This gazebo seems to defy all the rules when it comes to resisting the outward thrust of the roof. Built by Bullock & Company, an Ontario-based firm that specializes in round log homes, the gazebo has eight round log posts and eight long, heavy, round log rafters to match the lakeside home.

The posts are supported on piers rather than being sunk into the ground, and there are no ceiling joists to resist outward thrust. The weight of the massive center finial

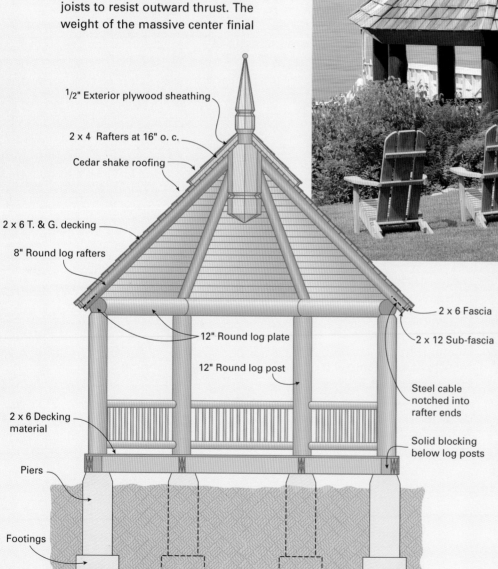

1/2" Exterior plywood sheathing

2 x 4 Rafters at 16" o. c.

Cedar shake roofing

2 x 6 T. & G. decking

8" Round log rafters

12" Round log plate

12" Round log post

2 x 6 Decking material

Piers

Footings

2 x 6 Fascia

2 x 12 Sub-fascia

Steel cable notched into rafter ends

Solid blocking below log posts

FIGURE 4

only serves to increase the pressure on the posts. The railings, which match those on the house, are mortised into the posts. While this does help resist outward thrust, it obviously is not enough to resist the force of that massive roof structure.

As shown in figure 4, the ingenious and simple solution devised by designer Tim Bullock is hidden in the roof overhang: The rafters have a notch in the end, and running through these notches is a steel cable that resists the outward thrust. The two ends of the cable are joined with a turnbuckle that was used to tighten the cable.

ceiling joists support a tower floor

In the gazebo we built for this book, 2 x 4 ceiling joists came together to support a platform that serves as the floor of the viewing tower. The eight-sided platform consists of tongue-and-groove boards with a blue-stained side facing down, the same material used for the main ceiling. An octagonal piece of plywood atop the tongue-and-groove boards formed the floor surface. Here you see only the plywood, which was temporarily screwed into place, to serve as a work platform until the tongue-and-groove boards were delivered.

The octagon was cut from a half- sheet of plywood, so the distance between opposing points was 4 feet. The layout was done in exactly the same way as laying out the gazebo itself, except we used a 4-foot square of plywood instead of a 10-foot-square piece of ground (see figure 2 on page 47).

Rather than use two through joists with a lap joint, we used one through joist and two half-through joists, mirroring the construction of the deck frame below. Plenty of screws through the tongue-and-

Blocks between joists stiffened the structure and supported the edges of the tower floor.

groove boards above made the connection strong. We drove screws through the side of the through joist into one of the half-through joists, and then we toe screwed the other half-through joist into place.

Next came four box joists, cut 22 inches long with 45° angles on the ends, just like the outer box joists in the floor. We attached the four remaining ceiling joists centered on the box joists.

Later, after the tower was framed, we added blocks between the ceiling joists that were underneath the platform, making the blocks flush to the edges.

With no sunken posts, and a massive roof without ceiling joists, this gazebo seems to defy the laws of physics.
PHOTO BY BRIAN VANDEN BRINK.

Ceiling joists meet at the middle to support the floor of the viewing tower.

Installing Top Plates

Before installing top plates, you need to cut the posts off level to each other. Because your level floor is already installed, all you need to do is measure from the floor to mark each post and then cut it off with a circular saw. The exact height of the posts will, of course, vary according to your design. You don't want tall people to have to duck into your gazebo, but unless your gazebo is very large you don't want disproportionately tall openings. As a rule of thumb, an opening height of about seven feet works well.

For a four-sided gazebo, top plates are installed just as they are in house framing. Use 2 x 4 plates for 4 x 4 posts or 2 x 6 plates for 2 x 6 posts. Measure the dis-tance from the outside of one post to the outside of another. Subtract the width of one post from this measurement. Cut eight pieces of 2 x 4 to this length, and preassem-ble them with an offset equal to the width of the plate stock, as shown in figure 5. Nail the assemblies to the posts and then to each other.

If you turn back and take an-other look at figure 1 on page 57, and figure 4 on page 62, you'll no-tice an important difference in the way the posts are oriented. In the double-beam plan, the posts are arranged in pairs parallel to every other rim joist. In the concentric plan, all of the posts face the cen-ter. As noted in the preceding chapter, six-sided gazebos always have a concentric plan.

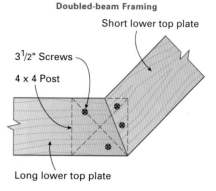

Doubled-beam Framing

Short lower top plate

$3\frac{1}{2}$" Screws

4 x 4 Post

Long lower top plate

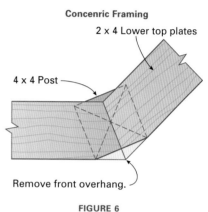

Concenric Framing

2 x 4 Lower top plates

4 x 4 Post

Remove front overhang.

FIGURE 6

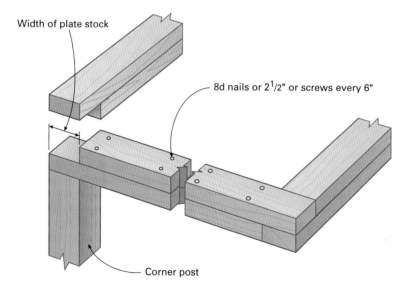

Width of plate stock

8d nails or $2\frac{1}{2}$" or screws every 6"

Corner post

FIGURE 5
Top Plates for Four-Sided Gazeb

1 **Cut and install the lower top plates.** Regardless of the fram-ing plan, the top plates for an eight-sided gazebo will be cut with $67\frac{1}{2}°$ ($22\frac{1}{2}°$ from 90°) cuts on both ends. However, in the doubled-beam orientation, the outside edges of the plates will meet at one corner of the posts, as shown in figure 6. As a result, the top plates whose front edges are flush to the outside of the posts will be longer than the plates that meet the posts at an angle. In the concentric fram-ing orientation, the plates are all the same length because they will meet across the center of the posts, as shown in figure 6. The plates land on the posts in the same way for a six-sided concentric plan; the

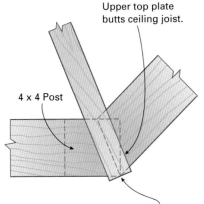

Upper top plate butts ceiling joist.

4 x 4 Post

End of ceiling joist, and/or bird's mouth plumb cut meets corner of post.

FIGURE 7
Joist Position for Double-Beam Framing

only difference of course, is that the plates will be cut at a 60° angle.

With a concentric framing plan, the top plates will overhang the fronts of the posts. Once all the top plates are in place, use a circular saw or a hand saw to remove this overhang.

2 **Cut and install the ceiling joists or rafters.** If you will include ceiling joints, install them now. Regardless of the orientation of the posts, the joists will land centered on the joints between lower top plates. Install the joists now, toenailing them to the lower top plates. If rafters will land directly on the posts, center the rafters on the plate joints and toenail them into place.

3 **Cut and install the upper top plates.** Cut the upper top plates to butt into the sides of the ceiling joists or rafters, as shown in figure 7. Toenail the upper top plates into place.

Installing a Header Beam

The header beam Jim Quinlan used on his gazebo consists of two 2 x 6s on end attached to a 1 x 6 soffit board below. To create a decorative detail, Jim routed a cove on the bottom edge of the soffit boards and then positioned the 2 x 6s so the soffit would have a ⁵⁄₁₆-inch reveal on both sides. (See figure 8.)

The decorative posts Jim Quinlan used for his six-sided gazebo are hollow, made of four 1½-inch-thick boards glued into a square and then turned. He left the posts square at the top and bottom. As mentioned, Jim notched the front of the posts at 60° to accept the outer 2 x 6 of the header beam. This system would work in exactly the same way for a solid, square 6 x 6 post. The soffit is not in the notch—it has its own notch to fit around the square portion of the post.

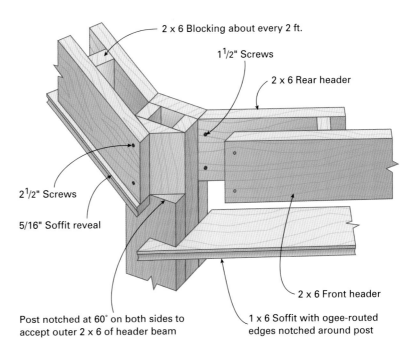

2 x 6 Blocking about every 2 ft.

1¹⁄₂" Screws

2 x 6 Rear header

2¹⁄₂" Screws

5/16" Soffit reveal

Post notched at 60° on both sides to accept outer 2 x 6 of header beam

2 x 6 Front header

1 x 6 Soffit with ogee-routed edges notched around post

FIGURE 8
Header Beam Notched Into Post

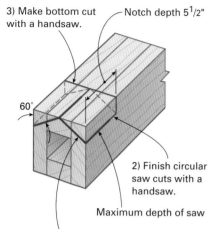

3) Make bottom cut with a handsaw.

Notch depth 5$\frac{1}{2}$"

60°

2) Finish circular saw cuts with a handsaw.

Maximum depth of saw

1) With circular saw base against end of post, and saw set to maximum depth, cut in from both sides at 60°.

FIGURE 9
Cutting the Post Notches

1 **Notch the post.** Following figure 9, lay out the 60° cuts on the top of a post. Next, lay out the depth of the opposing notches. Then lay out the bottom of the notches. Set your circular saw to maximum cutting depth and follow the 60° cut lines. Insert a handsaw into the circular saw kerfs and continue the cuts down to the bottom of the notch. Finally, use a handsaw to cut the bottom of the notch.

2 **Cut and install the rear headers.** Cut 2 x 6s with 60° angles on the ends to fit between the posts. The sides that face the inside of the gazebo should meet the inside corners of the posts. Attach the rear headers to the posts with 1½-inch screws that you drive into the face—the screws will be inside the header so they won't show.

3 **Cut and install the front header and blocks.** Cut 2 x 6s with 60° angles on the ends to the length you will need for the front headers. Set them in place in the notches and install with 2½-inch screws. Cut 1 x 6 blocks to fit between the front and rear headers, and screw them into place every 2 feet or so.

4 **Rout the soffit.** The routed reveal around the soffit is a nice optional touch. Jim Quinlan chose to rout a cove on the bottom of the reveal. I like the simple cove—it adds a nice detail without being too busy. You may choose a different profile, or you may leave it square or have no reveal at all.

5 **Notch one end of the soffit.** As mentioned, the soffit is notched to fit around the square part of the post. Before you cut each soffit board to its final length, cut a 60° angle on one end, as shown in figure 10. Lay out another 60° line parallel to the cut edge. Make the distance between the lines equal to half the width of the posts—for example, 2¾ inches for a 6 x 6. To allow for the front reveal, hook your measuring tape on the top edge of the soffit board and find the point where the 5⁄16-inch measurement intersects the 60° line. Draw a perpendicular line from this point to the cut edge of the board to complete the notch layout. Make the cuts with a circular saw, finishing at the corner with a handsaw.

6 **Notch the other end of the soffit.** To find the right length for each soffit board, start by measuring between posts from corner to opposing corner. Use this measurement to lay out the 60° cut that will butt into the side of the post. Find the 5⁄16-inch front reveal point and then measure over to a distance equal to half the post width. Draw a 60° line through this point to mark the end cut.

7 **Attach the soffit.** Use 8d galvanized finish nails to attach the soffit to the bottom of the header boards. There's no need to attach the soffit to the posts.

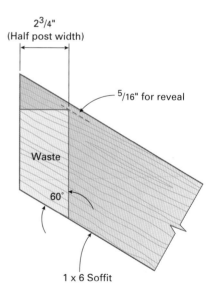

2³/4"
(Half post width)

5/16" for reveal

Waste

60°

1 x 6 Soffit

FIGURE 10
Laying Out the Soffit Notches

Installing Side Boards with Friezes and Brackets

If you capture the rafters with boards attached to the sides of the posts, you need to cut the posts off at an angle equal to the pitch of the hip rafters that land on the posts. To determine what this angle is, see Chapter 8.

As mentioned, the side board system is especially handy if you will be using friezes and and/or brackets, as in the case with the town park gazebo in Warwick, NY. Let's use that gazebo as an example of how the elements are assembled as units and then installed to the posts. (See figure 11.) Instructions for the railing are on page 109.

The first question to ask yourself is, how far should the friezes and brackets be from the top of the post? To answer this question, you need to plan the rafter overhang, especially if you plan to cover the ends of the rafters with a fascia board. In our example, the friezes are tucked up under the overhang a bit, even though the frieze top rails are 6 inches below the top of the side board.

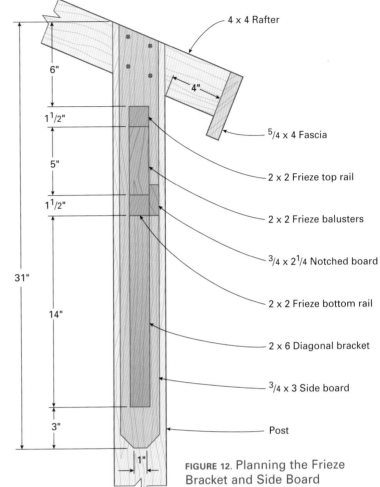

FIGURE 12. Planning the Frieze Bracket and Side Board

4 x 4 Rafter
5/4 x 4 Fascia
2 x 2 Frieze top rail
2 x 2 Frieze balusters
3/4 x 2 1/4 Notched board
2 x 2 Frieze bottom rail
2 x 6 Diagonal bracket
3/4 x 3 Side board
Post

6"
4"
1 1/2"
5"
1 1/2"
31"
14"
3"
1"

FIGURE 11. Frieze, Bracket, and Sideboard Assembly

2 x 2 x 5" Balusters
2 1/2" Screws
3" Screws
2 x 2 Bottom rail
1 1/2" Screws
3/4" x 2 1/4" Notched band
3/4" x 3" Side board
3"
5 1/4" R.
45
2 x 6 Diagonal bracket

Draw a side elevation sketch, similar to the one shown in figure 12, sketching in all the elements. Now plan out your railing heights as described in Complying with Building Codes on page 100, and draw an elevation of the whole post, railing, frieze, and bracket system as shown in figure 13. Do you like the proportions and the way views will be framed? If not, play with the length of the frieze balusters or width of frieze boards. Try raising or lowering the friezes or maybe eliminating the brackets. You might even want to go out to the gazebo and tack boards across the posts to help envision your design.

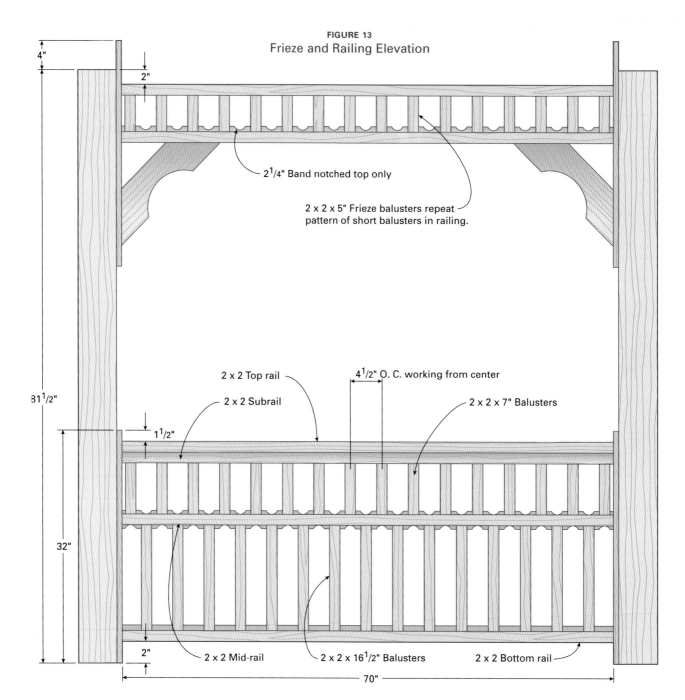

FIGURE 13
Frieze and Railing Elevation

4"

2"

2¼" Band notched top only

2 x 2 x 5" Frieze balusters repeat
pattern of short balusters in railing.

81½"

2 x 2 Top rail

2 x 2 Subrail

4½" O. C. working from center

2 x 2 x 7" Balusters

1½"

32"

2"

2 x 2 Mid-rail

2 x 2 x 16½" Balusters

2 x 2 Bottom rail

70"

1 **Make the side boards.** In this
design you make the side boards
from 1 x 4 stock ripped down to
3 inches wide to leave a ¼-inch post
reveal on both sides. Make the boards
long enough to go a few inches past
the rafter tops—you'll cut them off
in place after the rafters are installed.
If you like, make the decorative
angled cuts on the bottom.

2 **Cut the frieze rails.** Make the
rails and balusters of 2 x 2
stock. Cut the rails to the length
you will need with 22½° (from 90°)
cuts on the ends for an eight-sided
gazebo or 60° cuts for a six-sided
gazebo. If you are building the
eight-sided doubled-beam plan,
you probably realize by now that
four of the sets of rails will have

90° cuts on both ends, while four
will have 45° cuts. Note that the
rails and balusters are set back
from the front of the side board by
a distance equivalent to the thickness of the notched band. Be sure
to take this into consideration
when you measure for the rail
lengths—distances get shorter as
you move back on the angled posts.

3 **Cut the balusters and assemble the friezes.** Each frieze baluster is 5 inches tall. Cut enough balusters to space them 4½ inches on center. While you are at it, cut yourself a 3-inch length of 2 x 2 for use as a distance gauge for assembling the balusters to the rails. Mark the centerline of the rail width, and center a baluster on that line. Install the balusters working from the center out in both directions-the space between the end of the rails and the center of the last baluster at each end will be something less than 3 inches. Assemble the balusters to the rail using 3-inch wood screws in each end.

4 **Make the notched band.** More than a decorative element, the notched band prevents the balusters from turning and makes the whole frieze much stiffer. Make the bands from 1x stock ripped to 2¾ inches wide. Size and space the rounded notches to fall in the space between balusters. Cut the stock to the lengths you need with the appropriate angles on both ends.

Even though the rounded notches have a radius of ⅞ inch, you only cut them to a maximum depth of ½ inch from the edge of the board. To provide a surface for your compass's pivot point, put another piece of 1x stock alongside the band you are laying out, as shown in figure 14. Space the notches 4½ inches on center so that one notch will fall in the space between two balusters. Cut the notches with a saber saw.

When assembling balusters to rails, use a 3-in. piece of 2 x 2 as a spacer.

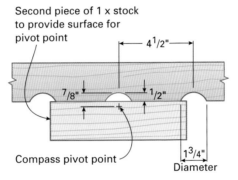

Second piece of 1 x stock to provide surface for pivot point

4¹/₂"

⁷/₈" ¹/₂"

Compass pivot point

1³/₄"
Diameter

FIGURE 14
Laying Out the Notched Band

5 **Assemble the friezes.** Position the notched board flush to the front of the side boards, and fasten each end by driving two 2½-inch screws through the side board and into the end of the notched board. Put the rail-and-baluster assembly in place, and drive two screws through the side board into the ends of the rails.

6 **Make and attach the brackets.** Make the brackets from 2 x 6s. Cut the pieces to 20 inches long with a 45° angle at the top. At the bottom, the bracket needs to be cut at a compound angle to meet the angled side board. To accomplish this, lay out a 45° angle on the face, and set a circular saw blade to the angle you need to meet the post. The rounded cutout has a radius of 5¼ inches, but only make it to a depth of 2 inches from the side of the board. As with the notched boards, use another piece of stock to provide a pivot point for your compass as you draw the arc. Cut the arc with a saber saw. Fasten the brackets to the bottom rail with 3-inch screws and to the side board with 2½-inch screws. Use two screws at each connection.

7 **Install the assembly.** Mark the posts for the bottom of the side boards. Center the side boards across the posts, and attach them with 2½-inch screws. Two screws above the frieze, one between the frieze and bracket, and one below the bracket should do the trick.

adding headroom at the entrance

As figure 13 shows, the bottom of the frieze is only 73½ inches from the floor. If the frieze over the door was that low, some folks would have to duck to enter the gazebo. But making the posts six or eight inches taller would have made the gazebo look disproportionately tall. Two things were done to solve this problem. First, the gazebo floor was notched back at the entrance so you don't actually step up onto the floor until you are past the entrance frieze. Second, the frieze curves up at the middle. Here's how to make the curved frieze as shown in figure 15:

By recessing the step and arching the entry frieze, plenty of headroom was provided without making the gazebo awkwardly tall.
PHOTO BY DAVID SCHIFF.

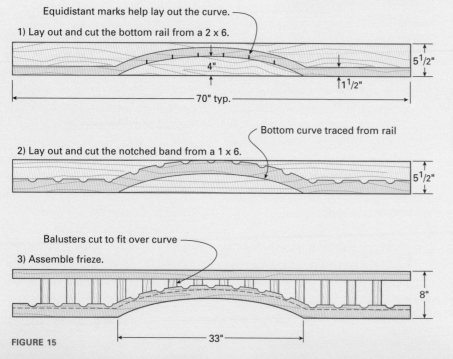

Equidistant marks help lay out the curve.

1) Lay out and cut the bottom rail from a 2 x 6.

4"   5½"
1½"
70" typ.

Bottom curve traced from rail

2) Lay out and cut the notched band from a 1 x 6.

5½"

Balusters cut to fit over curve

3) Assemble frieze.

8"

33"

FIGURE 15

1 **Cut the bottom rail and notched band to length.** Make the curved bottom rail from a 2 x 6 and the curved notched band from a 1 x 6. Cut both pieces to the length you need for the frieze, cutting angled ends to meet the posts.

2 **Lay out and cut the bottom rail curve.** Lay out the bottom curve of the frieze on the 2 x 6. Make the curve 33 inches wide and 4 inches deep. The exact shape of the curve isn't important—you can draw it freehand.

It helps to measure out some points equidistant to both sides of the curve's center. Plot as many points as you feel you need. Once you've laid out the curve, cut it with a saber saw. Sand out any irregularities.

3 Lay out the top of the bottom rail. Like all the rails, the curved bottom rail is 1½ inches thick. Use a straightedge to lay out the straight parts. Then set a compass to 1½ inches and run the pointed end against the bottom curve, allowing the pencil to draw the top curve. Cut out the top of the bottom rail, and sand it smooth.

4 Lay out and cut the notched band. Since the 2¼-inch-wide notched band has the same bottom curve as the rail, scribe the bottom curve from the rail. To lay out the top curve, set your compass to 2¼ inches. Cut and sand. Lay out and cut the notches as you did for all the other notched bands.

5 Assemble the curved frieze. Assemble the cut frieze in the same way as all the other friezes. Just cut the balusters to fit over the curve after you assemble all the other parts.

a variation on side-board construction

When last we left the gazebo built for this book, the 5/4- x 5-inch post side boards were running wild toward the sky, while the 2 x 4 in the middle of each post was cut to the height of the ceiling before the posts were assembled (see Post Construction, figure 6, page 67). The ceiling joists landed on the center 2 x 4s and continued past the posts to become nailers for the soffit boards under the roof overhang.

In this design, all the ceiling joists and rafters land directly on the posts. This means that the posts bear the entire downward load directly, and the ceiling joists resist any outward thrust. The only structural job left at the top of the posts is to prevent side-to-side movement. Rather than using plates or header beams, we simply used single 2 x 4s, angled at each end and cut to fit centered across the width of the post side boards. These were fastened in place with a couple of toe-screws at each connection.

Ceiling joists run through the built-up posts to become soffit nailers. A single 2 x 4 header prevents lateral movement.

CHAPTER SEVEN

Constructing Railings

Nothing, save perhaps the shape of the roof, states the style of your gazebo as much as your choice of railing. You can use straight, square balusters for a simple, elegant look, or choose turned balusters for a fancier effect. Solid boards provide some privacy when you are seated in the gazebo, while adding some cutouts to the boards will give you a Carpenter Gothic look.

Whatever style you choose, you'll build it easily using one of the systems described here. I'll start by describing how to put together a basic rail with square or turned balusters and show you how to adapt that design to boards or lattice. Next I'll show you how to build a railing that gives 2 x 2 balusters an attractive and practical twist. Then we'll move on to building the slightly more elaborate railing with notched bands, as shown in figure 8 on page 109.

For the grand finale, I'll show you how to make the sunburst railing featured in the gazebo plan at the back of this book. This railing is more challenging to make than the rest. But I've broken the process down into straightforward steps that certainly aren't beyond the skills of someone who has built that beautiful gazebo floor out there.

Building with Stock Rails

Lumber yards or home centers stock nicely shaped exterior rails with 2 x 2 square or turned balusters to match. These components come in either clear cedar or pressure-treated pine. While the cedar costs twice as much as the pine, it is worth every penny, even if you intend to paint the railing. Cedar is much more stable and much less likely to twist and crack than the pine—slicing pine into relatively thin railings and balusters is just inviting it to choose its own shape. Also, the pressure treatment serves little purpose since the railings are protected by the roof and won't stay wet long enough for fungus or other wood-eating critters to do any damage.

Stock rails have a ¼-inch-deep x 1½-inch-wide groove down their length. In the construction method shown here, you attach the top of the balusters to an infill strip and then screw the infill strip into the groove so you won't mar the top of the hand-rail with fasteners. For the bottom rail, the groove faces down, allowing you to just screw up through the groove into the bottom of the rails. The easiest way to attach these railings to the posts is to use side boards, as described on page 85.

1 **Make the infill strip.** Although the groove is only ¼ inch deep, I like to make the infill strip approximately ½ inch thick. I worry that a ¼-inch strip of wood, especially soft cedar, might allow the screws to break through, should the balusters suffer an impact. Rip the strips from a 2 x 4.

2 **Cut the rails and infill strips to length.** Fit the infill strips into the top rail stock. Cut all the top and bottom rails to the length and angles you need to fit between posts. Cut the infills at the same time as the top rails. Remember to allow for a ¾-inch-thick side board on each side.

3 **Lay out the baluster positions.** Simultaneously, with the strips still in their grooves, lay out the centerline and both sides of each baluster on the strips and bottom of the hand rail. Remove each strip and center it along the length of the top of the mating bottom rail. Scribe lines on the rail along both sides of the strip. Then transfer the balusters' positions to the bottom rail.

4 **Attach the balusters to the bottom rails and infill strips.** Drive one 2½-inch screw through the bottom of the bottom rail and into the bottom of each baluster. Drive two 1¼-inch screws through the infill strip into the top of each baluster. The second screw will prevent the balusters from turning.

5 **Attach the top rail and the side boards.** Fit the top rails over the infill strips and drive 1¼-inch screws through the infill strip into the top rail. Make sure to center one screw between the balusters. Using 2½-inch screws, attach the side boards to the rails at both ends of each rail.

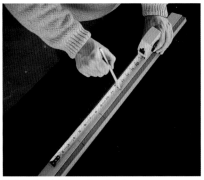
With the infill strips in their grooves, mark the strips and top rails for the baluster positions.

Transfer the baluster layout from the infill strips to the bottom rail.

Drive two screws through the infill strip into each baluster.

Screw the infill strips to the top rail.

positioning cap rails

Keep this is in mind when positioning railings across posts: When you cut a cap rail at an angle to meet a post, the angled cut will be wider than the rail. If the post is the same width as the cap rail, the cut will also be wider than the post. Since it is an aesthetic rather than structural decision, you can align either the front or the back of the cap rail to the edge of the post. (See figure 1.)

If you align the back, you'll probably want to saw the front of the railing flush to the front of the post, as we did for the gazebo we built for this book. In designs where you attach the railings before you attach the cap rail, a scrap of cap-rail stock cut to the correct angle will help you position the railings.

For the gazebo built for this book, we first positioned the cap rails flush to the back posts, then sawed the cap rails flush to the posts.

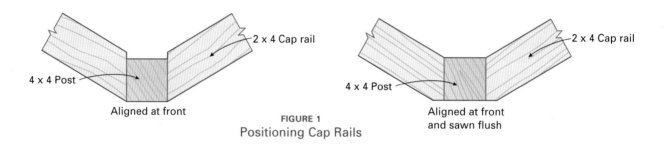

2 x 4 Cap rail

4 x 4 Post

Aligned at front

FIGURE 1
Positioning Cap Rails

2 x 4 Cap rail

4 x 4 Post

Aligned at front
and sawn flush

complying with building codes

Before designing railings for your gazebo, check with your municipal building department to see if your railing falls under local building code requirements. While requirements vary from town to town, codes usually require a railing if the floor of your gazebo is 30 inches or more above the ground. Codes also dictate the minimum height of the railing that can vary from 36 to 42 inches.

In addition, most codes have a maximum allowable distance between railing members to prevent small children from falling through or getting their head stuck. The most common space is 4 inches—in other words, you can't pass a 4-inch-diameter ball through the railing at any point, including the space between the bottom rail and the floor.

The All-Purpose Railing System

Here's a sturdy, straightforward, and easy way to build railings that you can easily adapt for square balusters, turned balusters, solid or cut boards, or even lattice panels. Its advantage is that rather than landing the balusters, boards, or lattice on a single bottom rail, you sandwich them between two side bands. This feature is particularly helpful if you live in a rainy location—water that collects on bottom rails can wick into the bottom of the balusters and potentially cause rot.

I'll start by showing you how to build this basic system using square or turned 2 x 2 balusters. Then I'll explain how you can adapt it for boards and for lattice.

You can make this railing in units, using side boards as previously described, or you can build it in place as described here. If you like, you can use trim-head screws for all the exposed connections.

While sandwiching the bottom of the balusters between bottom bands eliminates the need for a bottom rail, this gazebo uses one anyway for symmetry's sake.
PHOTO BY BETTY CROWELL.

building with balusters

1 Bevel and install the end balusters. The two end balusters in each railing are fastened directly to the posts. After cutting all the 2 x 2 balusters to the lengths you need, select two for each railing and rip a bevel along one edge to meet the angle of the posts. Determine whether you want the front or back of the cap rail to be flush to the post and position the end balusters accordingly (see figure 2). Make sure they are at the proper height, and fasten them every 6 inches using 3-inch screws.

2 Assemble the subrail to the balusters. Make a spacer block by cutting a piece of 2 x 2 to a length equal to the space between balusters. Cut the subrails to fit directly over the end balusters. Use your spacer block when attaching the subrail to the baluster tops. Use 3-inch screws to screw the subrail down onto the end balusters.

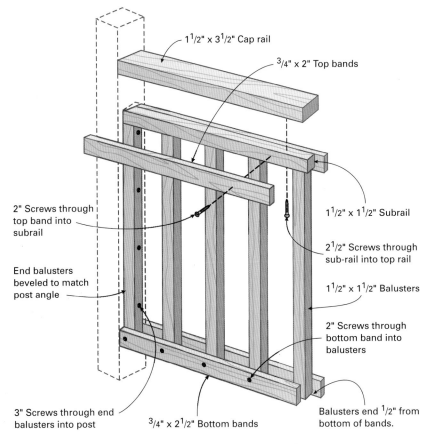

$1^1/2$" x $3^1/2$" Cap rail

$3/4$" x 2" Top bands

2" Screws through top band into subrail

End balusters beveled to match post angle

$1^1/2$" x $1^1/2$" Subrail

$2^1/2$" Screws through sub-rail into top rail

$1^1/2$" x $1^1/2$" Balusters

2" Screws through bottom band into balusters

3" Screws through end balusters into post

$3/4$" x $2^1/2$" Bottom bands

Balusters end $1/2$" from bottom of bands.

FIGURE 2. Building with Balusters

3 **Cut the bands and install the bottom bands.** Each railing has two ¾- x 2-inch top bands and two 2½-inch bottom bands. You can rip one of each from a 1 x 6. Cut the bands in sets of four to the length and angles you need to fit between posts. Remember, if you have angled posts, the outside bands will be a little longer than the inside bands. Now position the bottom rails to extend ½ inch below the balusters. Using 2½-inch screws, screw them to the end balusters. Use your spacer block when fastening each baluster to the bottom bands.

4 **Install the cap rail and the top bands.** Cut the cap rails to the lengths and angles you need. Attach them, using 2½-inch screws that you drive up through the sub-rail. Finally, put the top bands in place, and attach them to the end balusters and the subrail with 2-inch screws.

While you may not have the scroll saw required for the elaborate cut-board railing shown here, even making simple cutouts can greatly enhance your gazebo's style. PHOTO BY DICK DIETRICH.

The railing in figure 3 uses 5/4 boards, which are actually about 1⅛ inches thick, instead of balusters. The boards can be any width you like, and you can space them however you like as long as you comply with code. In the example shown in figure 3, the boards are 5/4 x 8 inches with ½-inch spaces between. The boards are notched to form 1½-inch square diamonds that are centered 5½ inches from the top and bottom. The diamond pattern is the simplest to lay out and cut. Of course, the cutout pattern you choose is limited only by what you can find to trace or copy or perhaps imagine and draw. Figure 4 on the next page provides a few possibilities to get your juices going.

Like the baluster design, you use a subrailing and sandwich the boards between the bottom and top bands. However, you don't fasten the subrailing to the top of the boards as with balusters, which, as you'll see, changes the order of assembly. You use cleats to secure the end boards to the posts and to provide bearing for the cap rail. Some board designs call for space between the end boards that precludes using cleats. You can eliminate the cleats by attaching the railings to side boards and attaching the side boards to the posts. Using hidden support blocks, as shown in figure 5, is a third method for attaching railings to posts.

1 **Cut and assemble the boards and bands.** Cut the boards to your chosen pattern. Cut the bands to the lengths and angles you'll need to meet the posts. Remember to take into account the final cap rail position as shown in figure 1 on page 100. Bevel the sides of the end boards to match the post angle. Fasten the boards to the bands using two 1¼-inch screws at the top and two at the bottom of each board. To leave space for the subrail, position the top band to extend 1 inch past the top of the boards. Position the bottom band to extend ½ inch past the bottom of the boards.

2 **Assemble the cap rail and subrail.** The subrail is made from a 1-inch-wide rip of 5/4 stock. Cut the cap rail and subrails to the length and angles you'll need. To prevent splitting, predrill the subrails every foot or so. Center the subrails onto the bottom of their mating cap, and attach them using 2-inch screws.

3 **Attach rails to the top band.** Insert the subrail into the space between the top bands. Attach the bands to the subrail using a 1½-inch screw approximately every foot.

4 **Cut and install the cleats.** Cut ¾- x 1¼-inch cleats to fit between the top and bottom bands. Bevel one ¾-inch edge of each cleat to the post angle. Predrill holes in the cleats about a foot apart for attaching the cleats to the posts. Predrill another set of holes about every foot on an adjacent side of the cleat for attachment to the end boards. Attach the cleats flush to the edges of the end boards using 1¼-inch screws.

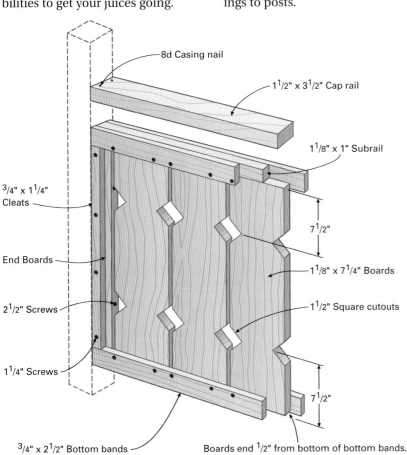

FIGURE 3. Building with Boards

5 **Attach the railings to the posts.** Typically, railings will be about 2 inches from the floor of the gazebo. Cut a couple of blocks to that height and place them against the posts. Set a railing on the blocks. Fasten the cleats to the posts with 2-inch screws through the predrilled holes in the cleats. Predrill and toenail 8d galvanized casing nails at an angle through the sides of the cap rail into the post and through the sides of the bottom band into the post.

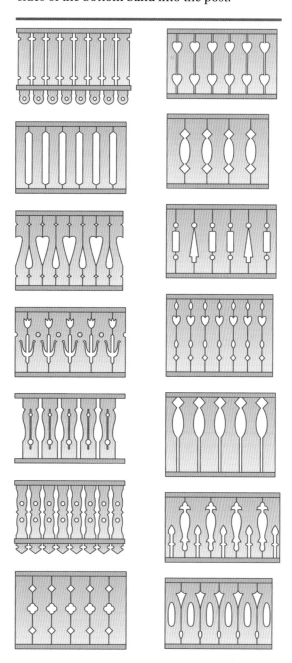

FIGURE 4. Board Designs

using hidden support blocks

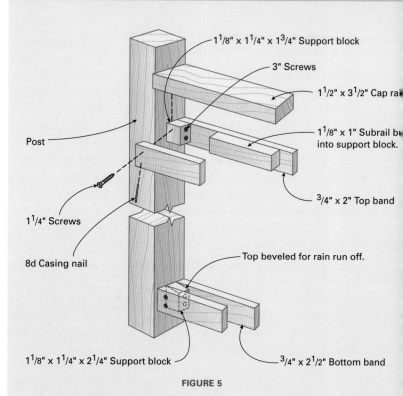

1⅛" x 1¼" x 1¾" Support block
3" Screws
1½" x 3½" Cap rail
1⅛" x 1" Subrail bu into support block.
Post
¾" x 2" Top band
1¼" Screws
8d Casing nail
Top beveled for rain run off.
1⅛" x 1¼" x 2¼" Support block
¾" x 2½" Bottom band

FIGURE 5

This method of attaching board railings to posts uses support blocks that eliminate the need for cleats or side boards. To make the support blocks, set your table saw to the angle of your posts and rip lengths of 5/4 stock to 1¼ inches wide. Cut enough stock into 1¾-inch lengths to make all the top support blocks. Predrill the blocks to prevent splitting, and attach a support block to each post at the proper height to support the bottom of the cap rail. Use two 3-inch screws at each connection.

Now assemble the board railing in the usual way, except make the subrail 1¼ inches shorter on each end to allow room for the support blocks. Test fit the railings in place over the top support blocks, and mark the posts at the top of the bottom band. Remove the railings. Cut more of your angled 5/4 rips to 2¼-inch lengths for the bottom support blocks. Cut the top of the blocks at an angle to allow water to run off. Screw the bottom blocks to the posts as you did the top blocks. (See figure 5.)

Finally, replace the railings, then fasten them to the support blocks using two 1¼-inch screws through both sides of each band.

Lattice comes in 4- x 8-foot panels. You can find pressure-treated pine or cedar panels at any home center or lumberyard. Panels of redwood and untreated pine are also available, but may be harder to find.

You can attach the sides and bottoms of lattice panels to the posts using cleats and a bottom band, as described in Building with Boards on page 103. However, since lattice panels are typically less than ½ inch thick, there isn't enough room between the top bands to fit a sturdy subrail. Instead of using the subrail, simply reduce the top band to the same 1¼-inch width as the post cleats.

You may also want to reduce the width of the bottom band to about 2 inches to keep it in scale. Just cut the lattice panels to size, and frame them with the bands and cleats on both sides. Because the lattice is thin, you can screw into one cleat or band, through the lattice, and into the cleat or band on the other side. Use 1½-inch screws driven from one side only. Predrill the cleats and the top band. Use 3-inch screws to attach the cleats to the posts and the top band to the bottom of the cap rail.

This simple gazebo with lattice rails and friezes creates a cozy garden nook. PHOTO BY CHARLES MANN, GARDEN DESIGN BY MARY SAMORA.

The railing on Jim Quinlan's gazebo is both beautiful and practical.

2 x 2 Balusters with a Twist

To me, the railing that Jim Quinlan designed for his gazebo is the perfect marriage of beauty and practicality. The gazebo's straight 2 x 2 balusters are turned 45° to the length of the rails. This not only makes the balusters more interesting to look at, it actually makes them stronger and a lot easier to paint. The bottoms of the balusters are notched to fit over the peak of a bottom cap rail. In addition to adding still more visual interest, this joint locks the balusters while the peaked bottom rail sheds water, preventing rot. The bottom cap rail overhangs the bottom rail, protecting that piece from the weather as well. (See figure 6 on page 107.)

Like other designs we've discussed, Jim's railing has a subrail screwed into the top of the balusters and into the bottom of the subrail. The top cap rail overhangs the subrail and, like the bottom cap rail, it has a peak to shed water.

Instead of being hidden behind a band, Jim celebrated this subrail with a routed cove along its length. The same cove graces the bottom of the bottom rail, tying together the railing's classical look.

And all of this is achieved with a few simple cuts on 2 x 2s, 2 x 4s, and 1 x 6s available at any lumberyard or home center. Here's how to do it.

1 **Cut stock for the bottom cap rail and bottom rail.** Make both the bottom cap rail and bottom rail from 2 x 4s ripped down to 2½ inches wide. Begin by ripping enough stock to make all the rails.

2 **Make the baluster cutting cradle.** To make the angled notches on the bottom ends of the balusters, you need to position them on edge while passing them through the blade of your table saw. Doing this safely and accurately requires a simple V-grooved cradle to hold the baluster during

cutting. You make the cradle from a piece of 5/4 x 6 inch or a 2 x 6 by ripping the stock down to about 4 inches, then cutting it to an 18-inch length. Crank the blade of your table saw to 45°, and raise it to cut through all but about ¼ inch of the cradle's thickness. Set the fence so the top of the cut will be roughly centered across the end of the cradle. Make the cut. Flip the piece around, and realign the fence so the top of the second cut will meet the first cut. Finish cutting the groove, then screw the cradle to the miter gauge on your table saw.

The second 45° cut meets the first to form the V-groove in the cradle that will hold the balusters during cutting.

3 **Set the saw blade to notch the balusters.** The angle at which the baluster notches meet the bottom cap rail is an aesthetic choice. Jim chose 60°—a familiar angle to someone who is building a six-sided gazebo. Set the blade to the angle you choose. Make the notch by placing a baluster in the jig, make a cut, then turn the baluster to 180° to make the second cut. Since you want the cuts to meet perfectly, you may have to run a few tests and make some adjustments to find the right blade height. To get in the neighborhood, draw a diagonal line across a baluster, place the baluster in the cradle, then raise the blade to the line.

4 **Notch the balusters.** The easiest way to make repeated cuts of the same length on the table saw is to clamp a stop block onto the rip fence. To work safely, the block must be behind the blade far enough so the work piece is no longer touching it while the cut is being made. Jim wanted the apex of his notches to be 24½ inches from the top of the balusters. So he set the block 24½ inches from the top of the blade. With the balusters in the cradle, cross cut all the notches.

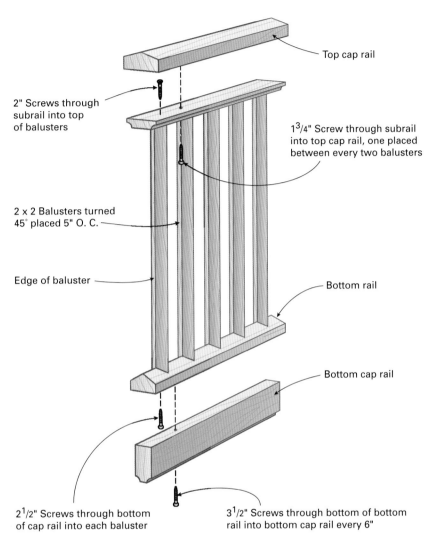

2" Screws through subrail into top of balusters

Top cap rail

$1^3/4$" Screw through subrail into top cap rail, one placed between every two balusters

2 x 2 Balusters turned 45° placed 5" O. C.

Edge of baluster

Bottom rail

Bottom cap rail

$2^1/2$" Screws through bottom of cap rail into each baluster

$3^1/2$" Screws through bottom of bottom rail into bottom cap rail every 6"

FIGURE 6
Rail Assembly

A diagonal line across a baluster helps set the saw blade height for the baluster notch.

A stop block clamped to the rip fence gauges the distance from the top of the baluster to the apex of the notch. For safety, the baluster must not touch the block during the cut.

Test cuts are made on a short piece of bottom rail stock to make sure the fence is positioned so the bevels meet at the exact center.

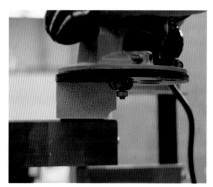

The bottom rail is clamped to the bench while a cove is routed onto both bottom edges.

5 Bevel the bottom cap rails. While you have your saw blade set to the angle for the baluster notches, use it to create the matching peak on the bottom cap rails. On one end of the board, center a mark across the width of the board. From that point, draw a line to each side of the board at the angle you need. Figure 7 shows the layout for 60°. Set the table saw fence and blade height to cut along the layout lines. To make sure your fence is set to the right distance, make a test on both sides of a short section of the bottom-rail stock that you ripped to 2½ inches. When the 60° cuts meet exactly at the center, cut the peaks on all the bottom cap rails.

6 Bevel the top cap rails. The top cap rails are 2 x 4s with peaked tops at a 15° bevel. This angle sheds water well, while still being gentle enough for a comfortable grip. Begin by laying out the beveled cuts, as shown in figure 7. Adjust the saw blade to 15°, and make the cuts just as you did for the bottom cap rail.

7 Make the subrails and rout the bottom rails. Make the subrails by ripping down 1 x 4s to 3 inches wide. Jim routed the bottom edges with a ⅜-inch-diameter cove bit set to a depth of ⁵⁄₁₆ inches after ripping the stock. As you can see in figure 6, a 1½-inch-thick side of the bottom rails meets the bottom cap rails. As Jim did, rout the bottom edges of the bottom rail using the same router bit and depth setting he did.

8 Cut the rails to length. Cut sections of top and bottom cap rails and bottom rails to the lengths and angles you will need. If you like, you can screw the subrails to the top cap rails and cut them as a unit—the bottom cap and bottom rail are too tall for that and need to be cut separately.

9 Assemble the railing. Mark the subrail and the bottom cap rail for the center of each baluster. Attach the subrail using one 1¾-inch screw in the top of each baluster. Next, attach the bottom cap rail using one 2½-inch screw in the bottom of each baluster. Attach the bottom rail to the bottom cap rail with one 3½-inch screw every 12 inches. Attach the top cap by driving 1¾-inch screws through the bottom of the subrail. Use one screw between every two balusters.

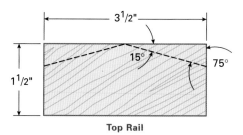

Top Rail

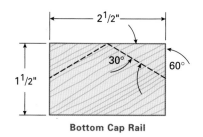

Bottom Cap Rail

FIGURE 7
Laying Out Cap Rail Bevels

Making a Railing with Notched Band

This railing is designed to accompany the frieze system described in Installing Side Boards with Friezes and Brackets, on page 93. The railing is constructed in essentially the same way as the frieze, so after reading the frieze section there will be just a few additional points to note.

Start by assembling the bottom rail and mid-rail to the 16½-inch-long balusters. Space the balusters in exactly the same way as the frieze balusters. Next, screw the 7-inch short balusters to the mid-rail, centering the short balusters between long balusters. Screw the subrail to the short balusters.

The railing has three bands. The top band is notched on the bottom only, and the middle band is notched at top and bottom. The bottom band is not notched at all. Just like the frieze, all the railing band notches are centered between balusters. This means the notches on the middle band are staggered top and bottom as shown in figure 8. Make the notches as described in step 3 of Installing Side Boards with Friezes and Brackets on page 93.

Attach the bottom band flush to the bottom of the bottom rail, the middle band flush to the bottom of the midrail and the top band flush to the top of the subrail. Center the cap rail over the subrail. In each

space between balusters, drive one 2½-inch screw through the subrail into the bottom of the cap rail.

Like the frieze, the railing is fastened to side boards, which are then attached to the posts. The side boards have decorative angled cuts at the top to match the cuts on the bottom of the frieze board.

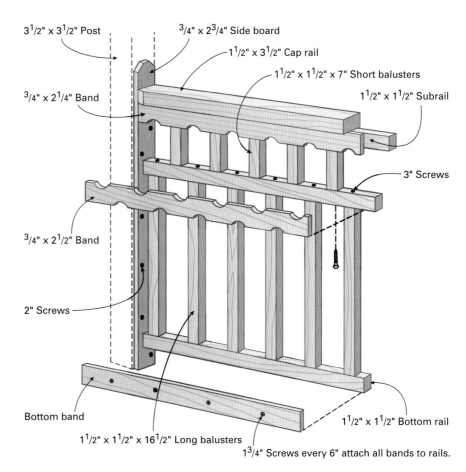

3¹/2" x 3¹/2" Post

3/4" x 2³/4" Side board

1¹/2" x 3¹/2" Cap rail

1¹/2" x 1¹/2" x 7" Short balusters

1¹/2" x 1¹/2" Subrail

3/4" x 2¹/4" Band

3" Screws

3/4" x 2¹/2" Band

2" Screws

Bottom band

1¹/2" x 1¹/2" Bottom rail

1¹/2" x 1¹/2" x 16¹/2" Long balusters

1³/4" Screws every 6" attach all bands to rails.

FIGURE 8
Railing with Notched Band

🏮 Making a Sunburst Railing

Let me start this section with a confession. When we set out to build this sunburst railing for the gazebo we built for this book, we really weren't sure how to do it. How should the rays relate to each other? What should the taper be? After much head scratching and a few false starts, we came up with a system that is surprisingly easy to follow. Now you can enjoy the "How did you do that?" compliments without having to go through the head scratching.

Just in case you were feeling guilty that we've done all the work for you, don't worry; you are still going to earn your accolades. I've broken the process down into steps, and while no step is difficult, there are plenty of them. The steps are organized into three sections: Laying Out the Sunburst Railing,

Cutting the Rays, and Assembling the Railing.

The number of rays in your sunburst and the dimensions of those rays will vary depending on the number of sides your gazebo

has and the length of those sides. However, since every single step in the three sections will still apply, you should have no problems with making the adjustments you need. Here we go!

This dazzling sunburst railing is not difficult to make as long as you break the process down into organized steps.

laying out the sunburst railing

Three measurements drove the design of this sunburst railing pattern. The first two were code requirements: Railings must be 36 inches high measured from the floor, and they must be designed so that a 4-inch-diameter ball can't fit through them at any point. The third measurement was 4½-inches—the width of the 5/4 stock we used to make the sunburst rays.

The easiest way to lay out your sunburst pattern is at full scale on a piece of luaun plywood or cardboard. As shown in figure 9, you

only need to lay out half of the railing. The other side is a mirror image, which allows you to just flip over identical boards.

1 **Lay out a post and the top and bottom rail.** Draw a vertical line to represent the side of one post and another vertical line to represent the center of the railing. Then draw horizontal lines to represent the top rail and bottom rail. Draw another horizontal line 1 inch from the bottom of the top rail. This is the top cutting line as shown in figure 9.

2 **Cut the first ray.** Begin by cutting a piece of 4½-inch-wide stock to 31 inches, which is the distance from the bottom of the bottom rail to the top cutting line—we tapered it down to 1¼ inches at the bottom because we liked the look of the angle on the taper; also, we knew that the angled rays would get narrower at the bottom, and we wanted to be sure we had enough width for attaching them to the bottom rail. Use a circular saw to cut the first ray. You'll use the first ray as a template to lay out the remaining rays.

3 **Lay out the first ray and first space.** Arrange the rays in pairs with a ¼-inch space between mates. Locate the first ray ⅛ inch from the centerline. This will make it ¼ inch from its mirror image mate on the other side of the railing. After drawing the first ray, measure 4½ inches along the top cutting line and make a mark for the location of the second ray.

4 **Lay out the second and third ray.** On the line for the bottom of the bottom rail, make a mark ¼ inch in from the first ray. Place the ray template on the layout as shown by the shaded areas on the drawing. Align its left side to the marks on the top cutting line and the bottom line. Align the top right corner of the template to the top cutting line. Trace the sides of the template to lay out the second ray. You'll need to extend the lines to the bottom line. Now repeat the process for the third ray, locating it ¼ inch in from the second ray. Use a strip of ¼-inch plywood to gauge the distance. Save the strip for use as a spacer when you install the rays.

5 **Lay out the fourth and fifth rays.** The only difference in laying out the fourth ray is that you align the top left edge to the top of the top band instead of the top right edge to the top cutting line. This is done to ensure good attachment to the top band. The fifth ray is the first one that doesn't reach the top band. For the fifth ray, position the top of the template along the same line as the fourth ray as shown in the drawing.

6 **Lay out the sixth ray.** Position the sixth ray simply as far away from the fifth ray as possible to a distance that will prevent a 4-inch-diameter ball from passing through—a can with a four-inch diameter makes a great gauge. Place the top left corner of the template on the post layout line.

7 **Lay out the corner piece.** For the seventh piece, place a length of the 4½-inch-wide stock ¼ inch from the sixth ray, and cut at angles to meet the post and the bottom line.

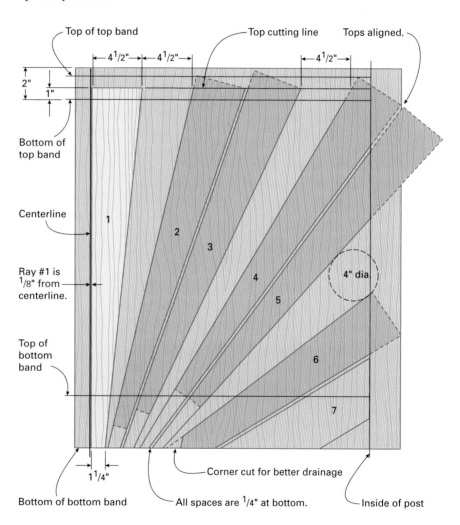

FIGURE 9
Laying Out a Sunburst Railing

cutting the rays

As figure 9 shows, our railing required seven different ray shapes. Each ray shape is used twice per rail section. That means our eight-sided gazebo required 16 rays in each shape for a total of 112 rays. The trick to cutting accurate tapers is to make a tapering jig. (See Making a Ray-Tapering Jig on page 113.) The key to organizing all this cutting—and keeping track of which ray is which during assembly—is to assign each ray shape a position number, as described in Laying Out the Sunburst Railing on page 110. Then, as you cut each piece, mark it with its position number and stack it in the appropriate position on the template.

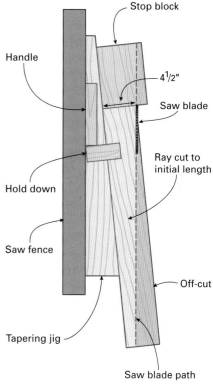

FIGURE 10
Setting the Saw Fence
for the Taper cuts

1 **Cut ray stock to initial length.** To get the initial length for each group of identical rays, take measurements from the full-scale layout. The initial length is the length of the untapered side from the top corner to the line indicating the bottom of the bottom rail. For example, in our layout the initial length of the first ray is 31 inches, and the second ray is 32¼ inches. Cut each numbered group to its initial length. Don't cut blanks for ray #7 though. Those short pieces don't get tapered, and you probably will be able to make some #7 rays from offcuts after all the other rays are made.

2 **Taper the rays.** Put the tapering jig against the table saw rip fence. Slide the jig forward until the leading edge of the stop block bumps into the saw blade. As shown in figure 10, set the saw fence so the distance between the blade and the jig's tapered edge is 4½ inches as measured along the stop block. Be sure to measure along the stop block's edge, not square to the blade. Note: The 4½ inches corresponds to the template width we chose; your measurement might be different. Now use the jig to cut the taper on all the rays. The first time you use the jig, you will cut off one end of the stop block.

3 **Cut the second and third rays.** You've already completed all the upright #1 rays. The rays that meet only the top bands are easy to cut. Just use a T-bevel on the full-scale layout to get the angles for the top and bottom cuts, and set the miter gauge on your table saw or the blade on your miter saw to the cut angle. Cut one of each ray shape to use as a template to lay out the cut angle, make your 16 cuts, then move on to the next angled cut.

The tapering jig allows you to quickly and safely taper all the rays with one setup.

4 **Cut the fourth, fifth, and sixth rays.** Make the bottom cuts for the rest of the rays in the same way as rays #2 and #3. Note that ray #6 has an extra cut on the bottom to help water drain through.

Rays #4 through #7 meet the post at a compound angle—22½° in the case of an eight-sided gazebo. This just means you have to set your table saw blade at 22½°. Ray #4 meets both the top rail and the post. The top cut is easy. The compound post cut was tricky because the steep angle combined with the tilted blade makes it difficult to hold against the miter gauge. The solution is to clamp the work pieces to the miter gauge.

Ray #4 meets the post at a steep angle that could have caused the work piece to slip during cutting. Clamping the piece to the miter gauge allowed the cut to be made safely and accurately.

5 **Cut the seventh ray.** As mentioned, ray #7 isn't tapered. Just cut the pieces to the angle to meet the bottom band and to the compound angle to meet the post.

making a ray-tapering jig

All of the rays in the sunburst railing have one square side and one tapered side. Since all the rays are laid out from the same template, they can all be cut with the same tapering jig. In fact, you can make the jig from the template ray itself, as shown in figure 11.

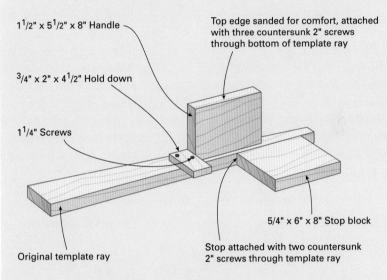

1½" x 5½" x 8" Handle

Top edge sanded for comfort, attached with three countersunk 2" screws through bottom of template ray

¾" x 2" x 4½" Hold down

1¼" Screws

5/4" x 6" x 8" Stop block

Original template ray

Stop attached with two countersunk 2" screws through template ray

FIGURE 11

1 **Make and attach the stop block.** For the stop block, cut a piece of the 5/4 ray stock to about 6 inches long. Place the stop block along the tapered edge of the template ray, making it an inch or two from the narrow end. Drive two countersunk 2-inch screws through the template into the block.

2 **Make and attach the handle.** The handle is a scrap of 2 x 8 approximately 8 inches long. For a comfortable grip, use sandpaper to round off the top edges of the handle. Put it in a comfortable spot on the template ray, positioning it flush to the non-tapered side. Use three countersunk screws to attach it through the template ray.

3 **Make and attach the hold-down.** The hold down will prevent the rays from rising up from the blade as you cut them. Use a scrap of ¾-inch wood or plywood that is about 2 x 4½ inches. Attach it against the handle with two 1¼-inch screws.

assembling the sunburst railing

Besides being pleasing to the eye, the sunburst railing is structurally very strong, since the rays are attached at several different angles. This helps stiffen the posts and adds rigidity to the whole roof structure, which is particularly useful in our gazebo because people will be climbing up into the viewing tower. With all rays cut, assembly is very easy. The whole railing is made of the same 5/4 cedar stock as the rays. We predrilled all screw holes, then used galvanized screws for all hidden connections, and stainless steel finish-head screws for all exposed connections.

1 **Cut and install the inner bands.** You will sandwich the rays between two sets of bands, with the top bands being 2 inches wide and the bottom bands being 4½ inches wide—the width of 5/4 stock. For each railing, start by cutting the inside top band top and the inside bottom band to the length and the angle you will need to fit between posts. Lay out the position of these bands so that the cap rails will be flush to the back of the posts. (See page 100, Positioning Cap Rails.) For each connection, use two 3-inch screws angled through the bands.

Notched scraps of plywood hang on the inside bands to act as spacers as the center rays are screwed through predrilled holes.

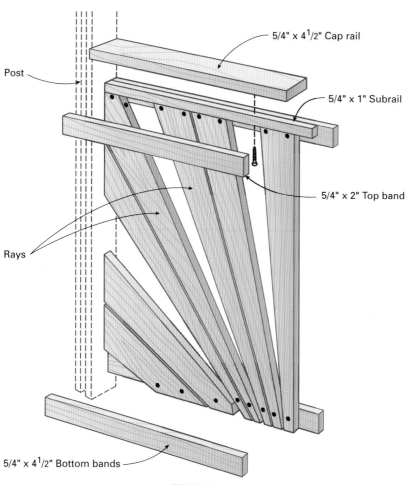

Post

5/4" x 4¹/2" Cap rail

5/4" x 1" Subrail

5/4" x 2" Top band

Rays

5/4" x 4¹/2" Bottom bands

FIGURE 12
Assembling the Sunburst Railing

2 **Install the center rays.** Take the strip of ¼-inch plywood used to lay out the rays, and cut it in two. Use a saber saw to notch the plywood scraps so they hang on the bands as you use them for spacers. Draw a horizontal centerline along the length of the top band. This corresponds to the top cutting line on your full-scale layout. Center the spacers on the top and bottom bands, and align the tops of the two #1 rays to the horizontal centerline. Secure the rays using two 1¾-inch screws with one at top and one at bottom, as shown in figure 12.

3 **Install the remaining rays.** Just follow the spacing on your layout to install the rest of the rays, continuing with two screws for each top connection and one screw for each bottom connection. Use 3-inch stainless-steel trim-head screws to attach the rays to the posts.

4 **Cut and attach the outside bands.** Cut the outside top and bottom bands at the angles and length you need to fit between posts. Attach them to the posts with 3-inch screws, then attach them to the top of the #1, #2, and #4 rays with 1¾-inch screws. Drive two 1¾-inch screws into the bottoms of the #7 rays.

5 **Assemble and install the cap rails and subrails.** Rip 5/4 stock for the subrails to a ¹⁵⁄₁₆-inch-wide cut, making them a little longer than you will need for fitting between posts. Cut the 2 x 4 cap rails to rough length. Center the 5/4 face of the subrails across the bottom of their cap rails, and attach with 2-inch screws in pre-drilled holes, placing them every 6 inches. Cut the cap rails and subrail assemblies to the exact lengths and angles you'll need to fit between posts. Fit each subrail into the space between the top bands. Drive 1¾-inch screws through the top bands every 6 inches, alternating between front and back.

The rays are attached to the posts with stainless steel trim-head screws. The plywood scrap gauges the distance between pairs of rays.

The subrail, already attached to the cap rail, is tapped into place between the top bands. A block of wood protects the cap rail from the hammer blows.

CHAPTER EIGHT

Building the Roof

Jim Quinlan loves math. He figured the rafters for his six-sided gazebo, cut them all on the ground, then saw his calculations confirmed as the rafters came perfectly together. "For me, that's the game," Jim said. "I enjoy seeing the numbers work."

Daniel Grant would rather skip the math and trust his eye. When he built the eight-sided gazebo featured in this book, he put up sample rafters, then stepped back to see how they looked. When he found the angles he liked, he

Jim Quinlan calculated and cut his rafters on the ground. It's a good thing numbers don't lie—Jim completed the gazebo just days before his daughter was married under its six-sided roof.

marked them in place. "I like to think with my hands and eyes first," Daniel said.

For Jim, numbers don't lie. Daniel would rather sketch in the sky. If you were building a house with 30-foot-long rafters, Jim's approach would be the only practical one. But depending on your design, construction method, and temperament, Daniel's approach can be perfectly practical for a whimsical structure such as a gazebo. After all, the roof only has to look good to you—it doesn't

have to follow some architectural drawing. I'll discuss both approaches in this section.

I'll start by giving you an understanding of roof geometry in general, then the geometry of gazebo roofs in particular. I'll tell you how to sheathe the roof and cover it with shingles or shakes. Along the way, I'll give you some ideas for applying these concepts to your own designs, including how to make kicked rafters and rafters that curve grace-fully along the top.

Understanding Roof Geometry

Ask a child to draw a house, and she almost certainly will draw the roof as a simple triangular gable. All of the rafters on a gable roof are called *common rafters*—they are positioned at 90° to the walls they land on and to the ridge board. Most carpentry books, logically enough, begin by explaining how to calculate and cut common rafters, then move on to discuss hip roofs. In a hip roof on a house that is

rectangular but not square, as shown in figure 1, *hip rafters* run from each corner of the building to meet a ridge board at a 45° angle. There are also hip jack rafters that run 90° to the top of the wall and meet the hip rafters at 45°. (It's a

little more complicated than that because the rafters rise as well as travel across the building, but we'll get to that later.) Hip roofs also have common rafters to fill out the sides.

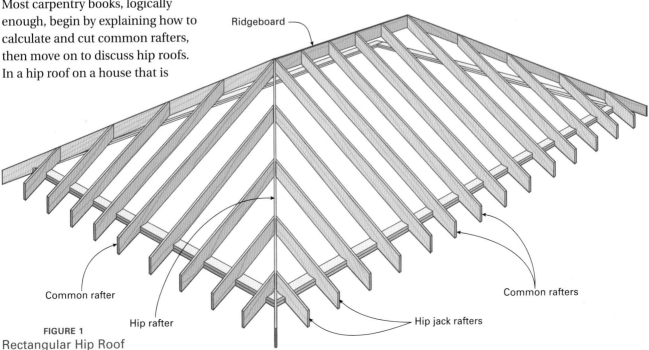

Ridgeboard

Common rafter

Hip rafter

Common rafters

Hip jack rafters

FIGURE 1
Rectangular Hip Roof

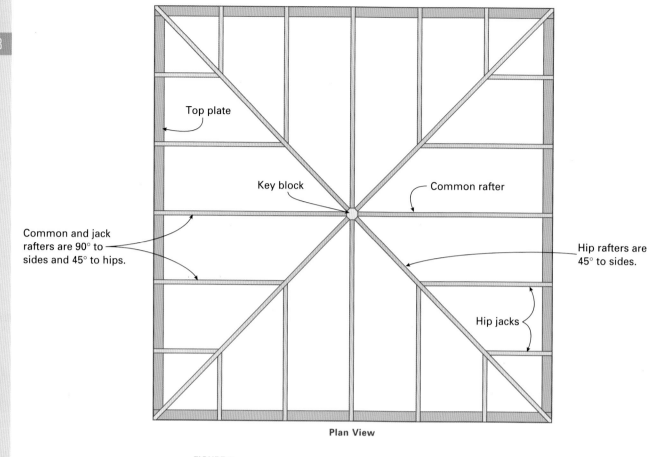

Top plate

Key block

Common rafter

Common and jack
rafters are 90° to
sides and 45° to hips.

Hip rafters are
45° to sides.

Hip jacks

Plan View

FIGURE 2
Framing a Square
Gazebo Roof

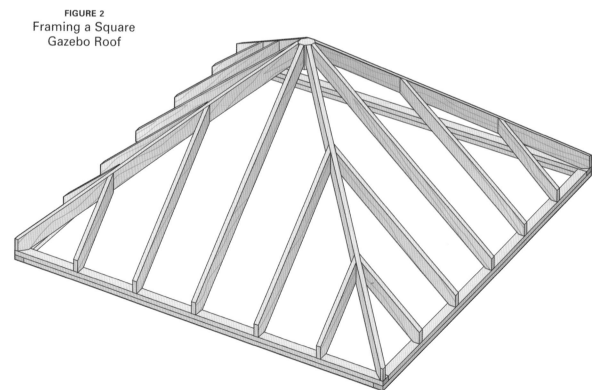

Like any rectangular building, a square gazebo will have four hip rafters. But because the roof is square, the hips converge in the middle, typically meeting at a key block rather than a ridge board, as shown in figure 2. In fact, if your gazebo does have four sides, all the information you need to calculate your rafters is right on your framing square—as you will see a bit later.

Of course, your gazebo will most likely have six or eight sides. This means that the hips will not be 45° to the sides. Instead, the hips will be at a 60° angle to the sides of a six-sided gazebo, or 67½ (22½° from 90°) to the sides of an eight-sided gazebo. Regardless of how many sides your gazebo has, the concepts for calculating the rafters will be the same—only the numbers will be different. And while your standard carpentry book or rafter square won't give you the numbers you need for six or eight sides, don't worry, I will.

The white rafters stand out against the beaded tongue-and-groove ceiling boards, celebrating the geometry of the hip and jack rafters in Jim Quinlan's six-sided gazebo.

choosing lumber dimensions for rafters

Several factors affect the lumber size you'll need for your gazebo rafters. The first considerations are the distance the rafters will span, and how far apart you will space the rafters. Other considerations include the lumber species and grade, and knowing if the rafters will be subjected to a snow load.

Luckily, unless your gazebo is the size of a two-car garage, you are unlikely to need rafter stock wider than 2 x 6. For example, structural grade Douglas fir 2 x 6s can span a sloping distance of 13 feet 3 inches when placed 24 inches on center. Reduce those rafters to 2 x 4 stock and they can span a sloping distance of 9 feet 1 inch. To measure the sloping distance, start at the inside edge of the top plate and measure to the point where the bottom of the rafter meets the key block.

An easy way to check if your rafter stock is beefy enough is to visit the Canadian Wood Council website at www.cwc.ca. There you will find SpanCalc, a program that lets you plug in your numbers to check allowable spans.

slope and the common rafter

Another unique factor about gazebo roofs is that while they always have hip rafters, they may or may not have common rafters and/or jack rafters. Some gazebos are small enough to allow the roof sheathing to span from hip to hip without intermediate rafters. Jim Quinlan's gazebo, for example, has two jacks meeting each hip, but no commons as shown in the photo on page 119. However, even if your gazebo won't have common rafters, it's important to understand how to calculate these rafters since they define the slope of the roof plane. You need to take into consideration that your roof sheathing will follow this slope whether or not there actually are common rafters below. To calculate the rafters you measure the slope and find the total rise.

Measuring the slope. In roof framing, slope is expressed as the number of inches of vertical rise per 12 inches of horizontal run. If a roof rises 6 inches for every 12 inches of run, carpenters will describe it as a 6 on 12 roof, or simply a six. A 6 on 12 roof rises at 22½°—you might find such a roof on a ranch-style house. A 12 on 12 roof, which rises at a 45° angle is common on Cape Cod style houses. Gazebos often have much steeper roofs—a 21 on 12 roof, rising at 60°, would not be uncommon.

Finding the total rise. Measure across the gazebo from the outside of one top plate to the outside of the parallel top plate, as shown in figure 3. This will give you the span of the common rafters. (Measuring across opposing posts would give you the span of the hips, but as you'll see, you'll never need this measurement.) Let's say we want to put a 9 on 12 roof on a gazebo with an 8-foot common span. To determine the total rise—the vertical distance from the top plates to the peak of the roof—first divide the 8-foot span in half to get the total run, in this case the total run is 4 feet. Since we know our roof rises 9 inches for every 1 foot of run, multiplying 9 x 4 tells us our peak will be 36 inches higher than the top plate.

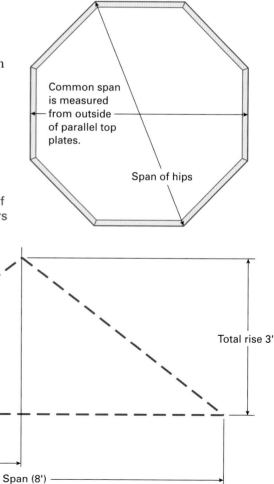

Common span is measured from outside of parallel top plates.

Span of hips

FIGURE 3
Rise and Run of
Common Rafters

Total rise 3'

Unit rise

12"
Unit run

Total run (4')

Span (8')

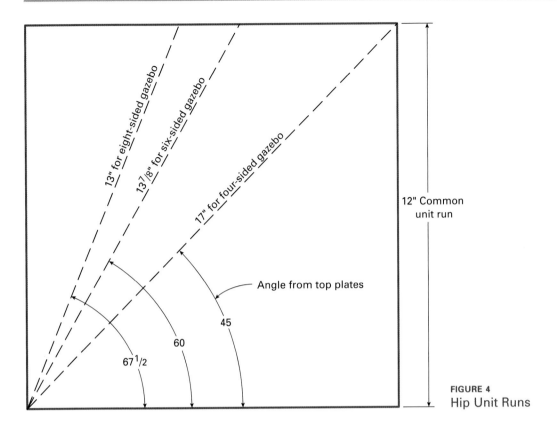

12" Common
unit run

Angle from top plates

45

60

67¹/₂

13" for eight-sided gazebo

13⁷/₈" for six-sided gazebo

17" for four-sided gazebo

FIGURE 4
Hip Unit Runs

If you draw a square with 12-inch sides and then a diagonal line between two corners, that line will be 16.94 inches long—or 17 inches, which is close enough for carpentry. With this fact in mind, take a look at the plan view in figure 2. When viewed in plan, the four common rafters divide the roof into four equal squares. The hip rafters span diagonally across these squares. This means that for every 12 inches of common rafter run, the hips will travel 17 inches. Given the common rafter slope is 9 on 12,

and our hip rafter slope is 9 on 17, what's the slope of the jack rafters? You guessed it! Since the jack rafters are parallel to the common rafters, the slope of the jack rafters is 9 on 12.

If you are building a six-sided gazebo, the hips will be 60° to the sides instead of 45°. At 60°, the hips will run 13⁷/₈ inches for every 12 inches of common run, as shown in figure 4. For an eight-sided gazebo, the hips will be at a 67½° angle to the sides, making them run 13 inches for every 12 inches of common run.

Calculating Rafter Length

So far we have been talking about rafters as if they were lines and points on paper instead of pieces of lumber with thickness and width. We have also ignored soffit overhangs. Figure 5 shows how the lines and points of the rise, run, and slope relate to a typical method of constructing rafters for a gazebo. As shown in the drawing, the rafter has a notch called a bird's-mouth to fit around the top and outside of the top plate. In the case of a hip rafter, the bird's-mouth sometimes sits directly on a post.

The line representing the slope runs from the outside of the top plate until it meets the line representing the total rise at a point we'll call the theoretical peak. This is known as the *line length* of the rafter. Because the rafter stock has width, the top of the rafter will meet the key block at a point above the theoretical peak. You don't have to consider it in your calculations—that's just something that happens. However, in your rafter-length calculations you do need to

allow for half the thickness of the key block, and also need to add in the length of the rafter overhang.

There are three ways to calculate the length of a common rafter: You can use the tables on your framing square, you can use the Pythagorean theorem, or you can use the framing square to *step off* the length directly on the rafter stock. You can use all of these methods to calculate hip rafters for a square gazebo. For a gazebo with six or eight sides, you'll need to use the Pythagorean theorem or the step-off method because the framing square doesn't have hip tables that apply.

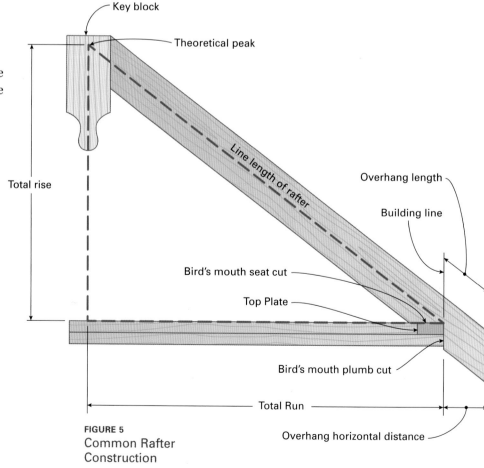

FIGURE 5
Common Rafter
Construction

using the rafter square table to find common rafter lengths

Along the length of the blade of your framing square you will find six lines of information about rafter lengths and cuts. The first line is labeled Length of Common Rafters per Foot of Run. Here's how to use that line to calculate the length of common rafters and hips for square gazebos.

1 **Find the rafter line length.** As an example, let's say your roof has a total run of 5 feet and a unit rise of 8 on 12. Look at the first line under the 8-inch mark on your framing square and you'll find that your rafters need to be 14.42 inches long per 1 foot of run, or, in this case, times 5 feet: 14.42 x 5 = 72.1 inches—we'll call that 72⅛ inches.

2 **Add the overhang length.** Now add in the length of the rafter overhang. If, for example, you want the overhang to protrude 15 inches horizontally from the plate, convert 15 inches to 1.25 feet. Then multiply 1.25 by 14.42 to get 18.025. This means you need to add 18 inches to your common rafter length: 72⅛ + 18 = 90⅛ inches.

3 **Subtract for key block thickness.** Let's say the key block is 5½ inches thick, making its portion of the total run 2¾ inches. A decimals-of-a-foot chart or construction calculator converts 2¾ inches to .2292 feet, then multiply: .2292 x 14.42 = 3.3046 inches. The table or calculator tells us we need to subtract 3⁵⁄₁₆ inches for the key block: 90⅛ - 3⁵⁄₁₆ = 86¹³⁄₁₆ inches, which is our final rafter length.

joining rafters to headers

Rather than using a key block to bring the rafters to a peak, your design may call for the rafters to meet headers. The headers, most often made of 4 x 4s to ease attachment and make them more stable, form a hexagon or octagon. The headers may support short posts that, in turn, support a second roof tier. Or, they may form a smaller hexagon to allow for a small cupola. (See Making a Cupola, page 136).

The Warwick Town Park Gazebo shown on page 82, sports both a second roof tier and a cupola. As shown in the photo, this gazebo has 4 x 4 rafters that meet short posts for the second roof tier. The assembly of the upper frieze, side short posts, and side boards mirrors the construction of the lower roof as described in Using Side Boards, page 85.

At the peak of the roof, the rafters meet eight short 4 x 4 headers that act as the base for the cupola. (See Making a Cupola, page 136.) The headers are cut at 22½° angles at both sides to form an octagon and are toenailed together with screws. To size the rafters to meet the headers, just treat the headers as if they were the key block when calculating the total run, total rise, and line length.

The construction of the bottom of the upper roof tier is a repeat of the method used below.

using the rafter square to find hip-rafter lengths for square gazebos

The second line on the framing-square table is labeled *Length of Hip* or *Valley per Foot Run*. To find the length of the hip rafters for our example roof with a total common run of 5 feet and a unit rise of 8 on 12, just look at the second line under the 8-inch mark. The number you'll find is 18.76. Multiply 18.76 x 5 to get a rafter line length of 93.8, or 93¹³⁄₁₆ inches. To get the 15-inch horizontal overhang, multiply 1.25 x 18.76 to get 23.45, or 23⁷⁄₁₆ inches. Add the 93¹³⁄₁₆-inch line length to the 23⁷⁄₁₆-inch overhang to get 117¼ inches. To subtract for the key block, multiply .2292 (or half the width of the key block in decimals of a foot) x 18.76 to get 4.2997, or 4⁵⁄₁₆ inches. Subtract 4⁵⁄₁₆ from 117¼ inches to get 112¹⁵⁄₁₆ inches. That means the hip rafters will be 9 feet, 4¹⁵⁄₁₆ inches long for our square gazebo with 86¹³⁄₁₆-inch common rafters.

using the pythagorean theorem to calculate common rafters

The Pythagorean theorem states that if you square the length of the base and altitude of a right triangle, the result will equal the square of the hypotenuse: $A^2 + B^2 = C^2$. Since the altitude describes the total rise of a roof, the base describes the total run, and the hypotenuse describes the rafter slope, we can use the theorem to determine rafter length. We'll use our same 8 on 12 roof with a 5-foot total run as an example again.

1 **Find the rafter line length.** Multiply the 5-foot run by itself to get 25 feet (A^2). Multiply 8 inches x 5 to find our total rise of 40 inches or 3.33 feet. Multiply 3.333 by itself to get 11.11 feet (B^2). Add 11.11 to 25 to get 36.11 feet (C^2). The square root of 36.11 is 6.009 (feet), or 72⅛ inches.

2 **Add the overhang length.** The overhang also forms a right triangle with a 15-inch horizontal distance as the base. That's a total run of 1.25 feet. Multiply 1.25 by itself to get 1.5625 feet (A^2). Multiply 8 inches x 1.25 to get a total rise of 10 inches or .8333 feet. Multiply .8333 by itself to get .6944 feet (B^2). Add 1.5625 to .6944 to get 2.2569 feet (C^2). The square root of 2.2569 is 1.502 (feet), or 18 inches.

3 **Subtract for key block thickness.** Again, we'll assume a 5½-inch-thick key block, making the block's total run 2.75 inches, or .229 feet. Multiply .229 by itself to get .05244 (A^2). Multiply 8 inches x .229 to get a total rise of 1¹³⁄₁₆ inches or .151 feet. Multiply .151 by itself to get .022 feet (B^2). Add .05244 to .022 to get .07444 (C^2). The square root of .0744 is .2728 feet, or 3¼ inches.

A line length of 72⅛ inches minus 3¼ inches (key block) equals a rafter of 86⅞ inches.

using the pythagorean theorem to calculate hips

By now you realize that the framing-square tables provide a less cumbersome way than the Pythagorean theorem to calculate common rafters and hips for square gazebos. The beauty of the theorem, though, is that it's applied in exactly the same way for all hips and commons no matter how many sides your gazebo has.

As an example, we've been using a gazebo with a 5-foot common-rafter run. As we know, the hips of a square gazebo have 17 inches of run for every 1 foot of common run. By multiplying 17 inches x 5, we find that the total run of the hips for our example will be 85 inches, or 7 feet,1 inch. Therefore, you can use exactly the same calculations as shown in the three steps above by plugging in a 7-foot, 1-inch rise instead of a 5-foot rise.

If your gazebo has six sides, the hips have 13⅞ inches of run for every 12 inches of common run. Multiplying 13⅞ inches x 5 tells you the hip run will be 69⅜ inches, or 5 feet, 9⅜ inches. For eight sides, multiply 13 x 5 to get a hip run of 65 inches, or 5 feet, 5 inches for our example.

With this method, you align the unit rise and unit run measurements on the framing square to the edge of the rafter square and *step off* each unit directly on the rafter stock. It's a great method for laying out rafters on site because it eliminates all the calculations. The biggest advantage is that the method is exactly the same for any common or hip rafter. The only thing that changes is the unit-run measurement on the framing square. As you already know, the unit run for all common rafters is 12 inches, and the unit run for hips on square gazebos is 17 inches. For six-sided gazebos, the hip unit run is 13⅞ inches, and for eight-sided gazebos it's 13 inches.

However, the step-off method is not very useful if you are making scale-drawing sketches on paper to help you envision your gazebo and roof. Another small drawback is that stepping off can be less accurate because any slight error you make in your marks accumulates—though this is less of a problem for short gazebo rafters than for long rafters used to build a house. Just be sure your pencil is sharp. When using the step-off method, or any rafter layout method for that matter, carefully lay out and cut one common rafter and one hip rafter, and then use them as patterns to cut the rest of the commons and hips.

As an example, let's use an eight-sided gazebo with a 9 on 12 roof with a total run of 5 feet, 6 inches. To lay out a common rafter on this roof, we would use the 9-inch and 12-inch marks on the framing square. By making our example a hip rafter, we'll use the 9-inch and 13-inch marks.

1 **Lay out the peak centerline.** Place the rafter across sawhorses. Because it's easier to manipulate the square, stand on the crown side of the rafter as you work, as shown in figure 6. The shorter, narrower leg of the framing square is called the tongue. The wider, longer leg is called the blade. Starting at the end of the rafter to your left (at top right in the drawing), align the 9-inch mark on the inside of the tongue with the edge of the board. Align the 13-inch mark on the inside of the blade with the edge of the board. Mark the peak centerline along the outside of the tongue.

2 **Lay out the odd unit.** The odd unit is simply the length left over after you divide the total run in inches by 12, making the odd unit in our 5 foot, 6-inch total run example 6 inches. The odd unit for hips must be adjusted. To do this, divide the odd unit by 12 (the common run). In this case $6 \div 12 = .5$. Now multiply the result by the hip unit run: $.5 \times 13 = 6.5$ To mark the odd unit, slide the square toward you, up on the centerline peak, and mark a distance of 6½ inches from the outside of the tongue to the crown of the rafter as shown in figure 6.

To give another example, let's say we want to calculate the odd hip unit for a six-sided gazebo with a total run of 5 feet 8 inches: $8 \div 12 = .67$. The hip unit run is 13⅞. Therefore, $.67 \times 13⅞ = 9⁵⁄₁₆$, which is the odd hip unit.

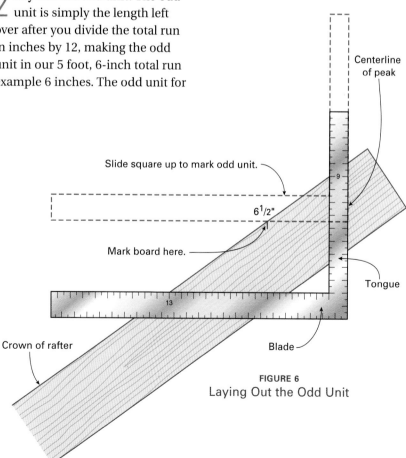

Slide square up to mark odd unit.

Centerline of peak

6½"

Mark board here.

Tongue

Crown of rafter

Blade

FIGURE 6
Laying Out the Odd Unit

3 **Step off the full units.** There will be one full unit for each full foot of total run—our example has five full units. To mark the first full unit, align the 9-inch mark on the inside of the tongue with the 6½-inch odd unit mark you just made, as shown in figure 7. Align the 13-inch mark on the inside of the blade with the crown edge of the rafter, and mark the unit. Now slide the square down to align the 9-inch mark on the inside of the tongue with the first unit mark and the 13-inch mark on the inside of the blade with the crown edge. Mark the second unit, continuing down until you've marked all the units.

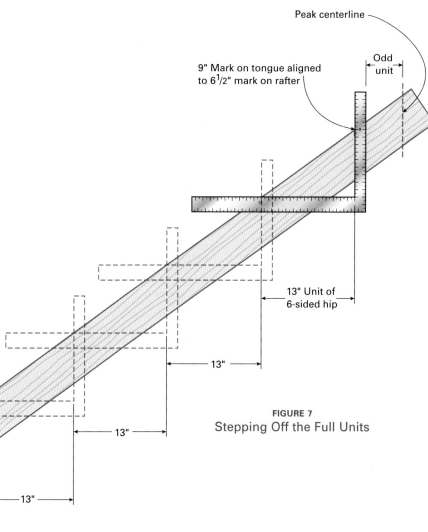

Peak centerline

9" Mark on tongue aligned to 6½" mark on rafter

Odd unit

13" Unit of 6-sided hip

13"

13"

13"

13"

Building line

13"

FIGURE 7
Stepping Off the Full Units

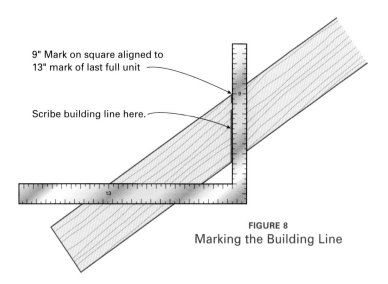

9" Mark on square aligned to 13" mark of last full unit

Scribe building line here.

FIGURE 8
Marking the Building Line

4 **Mark the building line.** The 13-inch mark of the last full unit marks the position of the outside of the top plate. This is known as the *building line*. Move the square down and, once again, align the 9-inch mark on the inside of the tongue with the last 13-inch mark on the rafter, then align the 13-inch mark on the inside of the blade with the crown edge. Scribe the building line on the rafter along the inside of the tongue, as shown in figure 8.

5 **Lay out the bird's-mouth.** You want the seat cut of the bird's-mouth to be as deep as the top plate is wide—3½ inches in the case of a 2 x 4 top plate. Go around to the bottom of the rafter and lay the outside of the tongue along the building line, as shown in figure 9. Slide the square up until you can align the measurement on the 3½-inch mark on the outside of the blade to the bottom edge of the rafter. Draw in the seat cut.

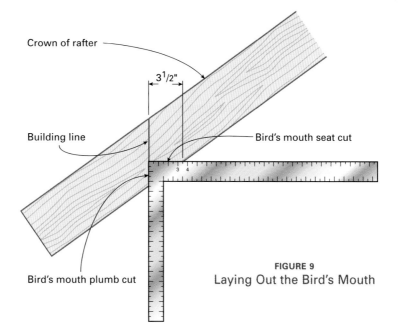

FIGURE 9
Laying Out the Bird's Mouth

6 **Lay out the overhang plumb cut.** Let's say you want an overhang of 15 inches as measured horizontally from the building line. This works out to one full unit of 12 plus an odd unit of 3. Since this is an eight-sided hip, start by laying out a full 13-inch unit run from the building line, as shown in figure 10. Draw a line on the rafter for the full overhang unit. Now you have to adjust the odd overhang unit. To do this, divide 3 by 12: 3 ÷ 12 = 0.25. Multiplying 0.25 by 13 gives us the

adjusted odd unit: 0.25 x 13 = 3¼ inches. Lay out and draw a parallel line 3¼ inches from the full unit line. This is the line for the overhang plumb cut. Alternately, you can make the cut square to the rafter instead of plumb.

7 **Lay out the top plumb cut.** To lay out the top plumb cut, you need to shorten the rafter to allow for half the thickness of the key block. To do this, just measure over and lay out the top plumb cut parallel to the peak centerline.

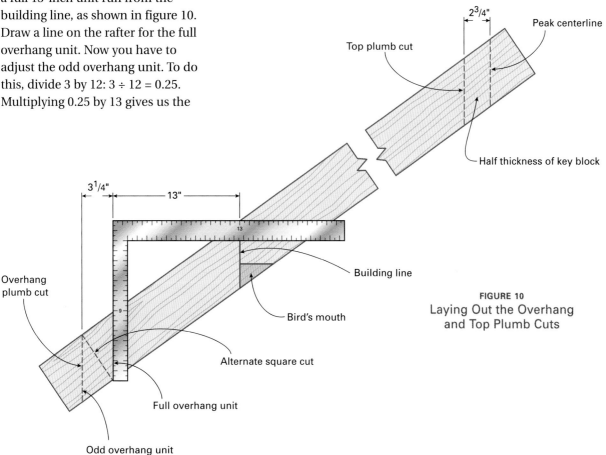

FIGURE 10
Laying Out the Overhang
and Top Plumb Cuts

laying out jack rafters

Because jack rafters are always square to the top plates, they follow the same slope as common rafters. This means that if you are building a 9 on 12 roof, the jacks will have 9 inches of rise for every 12 inches of run, regardless of how many sides the gazebo has. And, because the space between jacks is regular, the jacks will get shorter by the same amount as you get closer to a corner.

DETERMINING THE DIFFERENCE IN JACK LENGTHS FOR SQUARE GAZEBOS

The rafter square has tables for the difference in lengths of jacks. But as with hips, these tables are only useful for square gazebos. The third line on the framing square is labeled *Difference in Length of Jacks 16 Centers*. The fourth line is marked *Difference in Length of Jacks 24 Centers*. If you are not using these spacings, the rafter-square tables are useless even if you are building a four-sided gazebo.

For example, in the first step in Using the Rafter Square Table to Find Common Rafter Lengths for Square Gazebos (see page 124), we determined that the line length of a common rafter for an 8 on 12 roof with a 5-foot run would be 72⅛ inches. If our rafters will be 24 inches on center, we can check the fourth line under the 8-inch mark on the square. We'll find that the longest jack rafter will be 28⅞ inches shorter than the common rafter, and that each rafter will get 28⅞ inches shorter as we move toward the corners.

DETERMINING THE DIFFERENCE IN JACK LENGTHS FOR GAZEBOS WITH SIX OR EIGHT SIDES

Two factors affect the length of jack rafters: one is the angle at which they meet the hip rafters, the other is the slope of the roof. In this case, I can't give you any magic numbers to plug in. Truly, the easiest approach is to cut and install the hip rafters and then go up and take a couple of measurements.

Start by laying out lines on the top plates that represent both sides of each jack rafter. Then, to determine the length of the shortest jack, as shown in figure 11, hook your tape measure over the outside of the top plate at the line representing the side of the jack that will be farthest away from the meeting hip. Measure straight up to the hip. You might even want to have someone use the framing square to make sure the tape is perpendicular to the top plate when you take your measurement. Now measure for the second jack in the same way. If there are longer jacks, you can either measure for each length, or just increase each jack length by the difference between the first two jacks.

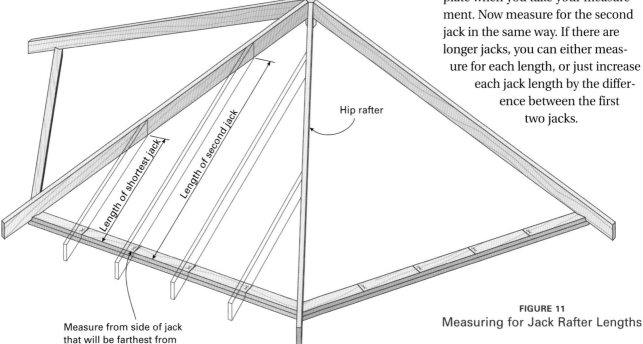

Hip rafter

Length of shortest jack

Length of second jack

Measure from side of jack that will be farthest from the hip it will meet.

FIGURE 11
Measuring for Jack Rafter Lengths

cutting jack rafters

The jack rafter cut meets the hip rafter at a compound angle, plumb to the side of the hip and angled to the face of the hip. You may be able to set up this cut on a compound sliding miter saw, but if you don't have such a saw, it's easy to make as two cuts with a portable circular saw and a simple support jig.

The jack rafter meets the hip with a plumb cut and a wide cheek cut.

1 **Lay out and make the top plumb cut, bird's-mouth, and overhang plumb cut.** Now that you know the length of the jack rafters, start by cutting the top plumb cut, the bird's-mouth, and the overhang plumb cut exactly as if the jacks were shorter common rafters. (See The Step-Off Method, page 125). Remember, since jack rafters always have a common run, you would use the 9- and 12-inch marks on the framing square for our 9 on 12 gazebo, no matter how many sides it has.

Make the top plumb cut on the jack rafter exactly as you would for any common rafter.

2 **Set the saw for the cheek cut.** The cheek cut creates the surface that will meet the hip rafter. Start by setting your circular saw blade at the angle that complements the angle of the hip rafters. The settings are 22½° for an eight-sided gazebo, 30° for six sides, and 45° for four sides. Set the blade to cut as deep as it will go.

making a cheek-cutting jig

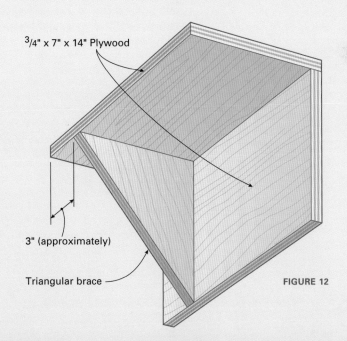

¾" x 7" x 14" Plywood

3" (approximately)

Triangular brace

FIGURE 12

You can make the jig used to stabilize the base of the circular saw during the cheek cut with just three scraps of ¾-inch plywood. Take two pieces that measure approximately 7 x 14 inches and screw them together as shown in figure 12. Then add a triangular brace about 3 inches from one end.

A simple jig set flush to the plumb cut provides a stable base as you make the cheek cut.

3 **Make the cheek cut.** While making the cheek cut, the saw base will run along the end of the rafter. You will need to clamp the rafter to a sawhorse with the crown down and the top plumb cut facing you at a comfortable height. Since the end of the rafter doesn't provide enough surface for a stable cut, clamp the cheek-cut jig to the rafter. Place the jig on the side opposite the one that will get the cheek cut, making sure the jig is flush to the end of the rafter. Make the cut.

4 **Finish with a hand saw if necessary.** Since steeper slopes require deeper cheek cuts, you may find that the circular saw blade will not make it all the way through the cut. If not, finish the cut with a hand saw.

It is often necessary to finish deeper cheek cuts with a handsaw.

sketching in the sky

When Daniel Grant designed the gazebo we built for this book, he only did a rough sketch of what he wanted it to look like. To Daniel, the fun of building a whimsical building like a gazebo is discovering its exact shape and style as he goes along. (As you probably have gathered by now, the detailed, dimensioned drawings didn't come until *after* the gazebo was built.)

Daniel's method of construction lends itself to this build-as-you-go approach. He likes to let boards run wild whenever he can, cutting them off in place as he makes aesthetic decisions. As you know, his posts for our gazebo consist of a center 2 x 4 flanked by 5/4 x 5 side boards. He precut the 2 x 4s to support 2 x 4 ceiling joists, while he allowed the 5/4 x 5 boards to run wild toward the sky. The 2 x 4 ceiling joists continued past the posts to support the soffit board that forms the bottom of the roof overhang. (See A Variation on Sideboard Construction, page 97.)

In fact, Daniel allowed the joists themselves to run wild, cutting them in place only after he decided how long the overhang should be. There's another advantage to cutting the joists in place: If there is any variation in the distances between the center of the gazebo and the posts, it won't affect the overhang lengths, which are measured from the outside of the posts.

Daniel's gazebo has two sets of rafters. The first set comes up to the bottom of the windows in the viewing tower. This raised a question. How high should the bottom of the windows be to provide safe enclosure while best framing the view? Since this would affect the slope of the bottom rafters, Daniel, guessing that it would give him the right slope, cut a 30° angle on the end of a 2 x 4. He climbed up on the tower platform, placed the angled end of the 2 x 4 between the post side boards and on top of the overhanging joist, then clamped this sample rafter to an upright piece of stock. Using a five-gallon bucket as a makeshift bench,

Daniel Grant enjoys a preview of what he'll see from the viewing tower of his gazebo while trying out a rafter slope.

Three screws through the side boards on each side attach the lower rafters to the tower posts. The two screws protruding from the side board were left loose to make it easier to slip the rafter into place.

With the lower rafters and tower posts installed, the next step was to add the 2 x 4s that form the lowest part of the windows.

Daniel got a good idea of what the view would be. Looking from the ground, we also got a good idea of the rafter pitch. As it turned out, a 30° angle on the bottom of the rafters worked just fine.

The viewing-tower posts are constructed just like the main posts, with the upper ends of the joists landing on the center 2 x 4 and the flanking 5/4 x 5 boards running wild. To lay out the angled cut for the top of the center 2 x 4s, all Daniel needed to do was make sure that the upright piece he had clamped to the rafter was plumb. Then he scribed a line where it met the bottom of the rafter.

After cutting the center 2 x 4s and the lower rafters to size, Daniel assembled the rafters to the posts on the ground. Just as the lower posts straddle the floor joists, he left space at the bottom for the side boards of the tower posts to straddle the ceiling joists. Then Daniel slipped these assemblies

over the ceiling joists and between the side boards of the main posts and screwed them in place.

Once Daniel had the lower rafters in place, he cut the main post side boards flush to the tops of the rafters. If we had thought of it, we would have angled the cuts slightly away from the rafters so the side boards wouldn't get in the way of the roof sheathing. However, we were able to make the angles later using a grinder.

Once Daniel installed the viewing-tower posts, the next step was to choose a slope for the tower rafters. Daniel started by cutting

45° plumb cuts on the tops and bottoms of a couple of sample rafters. He tacked them together with a couple of screws and a scrap crosspiece. He took this assembly up to the tower and clamped it in place. After looking at the gazebo from various vantage points around the property, the consensus was that the roof looked truncated— not dramatic enough. Daniel tried again. This time he used a steeper 60° angle, figuring that perhaps the angle that mathematically complements the 30° lower roof would complement it visually as well. All agreed that indeed it did.

A 45° upper roof angle didn't seem quite right, as shown in the photo on the left. As it turned out, a steeper, 60° angle complemented the 30° angle of the lower roof on the right.

Assembling the Roof

If you have stuck with me this far, you now have all the knowledge you need to lay out and cut rafters for any gazebo you might envision. Now, let's take a look at some ways to put the roof together.

I've already discussed how to lay out bird's-mouths. Cutting them is just a matter of starting two cuts with your circular saw, then finishing the cuts with a hand saw. In most cases, you toenail rafters through the bird's-mouths into the top plate, beam, or post. Or you can make the connection using hardware. (See Connectors for Six-Sided Gazebos, page 43.)

Now we're ready to move on to a discussion of key blocks, which you use whenever gazebo roofs come to a peak. If you are planning to use a cupola at the top instead of a key block, you can purchase a ready-made cupola or make one yourself as described on page 136.

To avoid over cutting the bird's-mouths, make two cuts with a circular saw until the cuts meet at the top, then finish the cuts with a handsaw.

making key blocks

Even a four-sided gazebo is likely to have eight rafters that converge at the peak—four hips and four commons. (See figure 2, page 118.) One way to join eight converging rafters is to use a four-sided key block. This allows four rafters to butt squarely to the four sides of the key block. The remaining rafters have 45° cuts on both sides to fit between, as shown in figure 13. If you like, you can angle the top of each side of the block to bring the roof planes to a point.

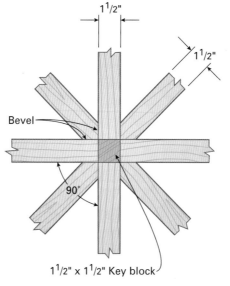

1¹/₂"

1¹/₂"

Bevel

90°

1¹/₂" x 1¹/₂" Key block

FIGURE 13
Square Key Block

OCTAGONAL KEY BLOCKS

More typically, eight rafters will meet an eight-sided key block, which is easy to make on many table saws. On most table saws, you can angle or tilt the blade to the operator's right. This means you have to make the cuts with the fence to the left of the blade. However, not all saws allow you to set the fence far enough to the left of the blade to make these cuts. In this case, you can make the cuts by clamping the key block to a bench and using a circular saw with the blade set to 22½°. To guide your cuts, use a rip guide with the circular saw or follow layout lines on the sides of the key block.

Make the key block from a length of 4 x 4. Lay out an octagon on one end using the same method of diagonal lines and points you used to lay out the gazebo posts. (See figure 2 on page 47.) Crank the saw blade over to 22½°. Make four cuts for an octagon.

Since all 4 x 4s do not measure exactly 3½ inches square, you may discover that the eight facets are slightly less than 1½ inches wide. If so, take a few licks with a block plane to taper the ends of the rafters slightly so they all fit against the block. A snug fit will make the roof structure a little stronger, anyway.

As mentioned, you may want to bring the top of the key block to a point, angling each facet of the key block to match the roof slope. Since the key block shown in figure 13 on page 132 will actually protrude from the roof and be covered by a metal weathervane, we simply nipped off the top of the facets at 45° to make it easier to fit the weathervane over it. However, the table saw procedure is exactly the same. Just lay out your cuts on one of the facets. Set the saw blade for the angle you need. Put the key block against the miter gauge and position it to make the cut. Clamp a guide block to the rip fence, and move the fence until the guide block touches the top of the key block, then lock the fence. To prevent dangerous kickback, make sure the key block is not in contact with the guide block during the cut.

Four cuts on the table saw turn a 4 x 4 into an octagonal key block.

A guide block clamped to the back of the fence gauges the position of the key block for the angled top cuts.

key blocks with six sides

To make a hexagonal key block, start by ripping a 4 x 4 to exactly 3 inches square. Draw diagonals across the end to find the center point. Set a compass to a 1½-inch radius, place the pivot point where the diagonals cross, and draw a circle. Center the pivot on one side of the block and, with the radius still set at 1½ inches, draw an arc through the circle. Set the pivot at the intersection of the arc and draw the next point, as shown in

figure 14. Continue around until you've laid out the six corners of the hexagon. At the table saw, make two opposing cuts with the saw blade at 90°, then set the blade to 30º for the remaining cuts.

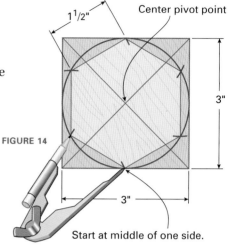

FIGURE 14

1½" Center pivot point

3"

3"

Start at middle of one side.

assembling the rafters to the key block

Rather than struggling atop the gazebo to keep the rafters aligned to the key block while driving screws, preassemble the roof frame on the ground. Then, unless you have a small crane handy, you'll need to disassemble them. When you reassemble the roof up top, alignment will be easy as the screws find their predrilled holes. When assembling the rafters both times, be sure to alternate sides of the key block as you do. If you work your way around the key block, the last rafters may not fit, especially if they are snug.

For the gazebo built for his book, we left the 2 x 4s out of the center tower posts until after the rafters were installed. This way, we could slip the rafters between the 5/4 boards that form the outside of the posts and screw them to the key block while the peak was still within easy reach. Then we pushed the rafters up to a premarked line on each post and screwed them in place.

The rafters for the lookout tower are preassembled on the ground. Each rafter gets one screw through the top and one through each side.

The rafters are attached to the key block while they are still in easy reach.

If you have wood-turning skills and a lathe, you may want to make a key block with a decorative turning at the bottom. For an octagonal key block, it's easiest to cut the eight facets first, then turn the decorative part. But, as you saw on page 134, cutting a hexagonal block on the table saw requires several fence adjustments. To forgo all that, you can make the six facets right on the lathe, using this clever method.

1 **Turn the block round, and mark the tape.** Start with a 4 x 4 that is as long as you need to meet the rafters, and make your decorative turning. Turn the block to a cylinder with a 3-inch diameter. Wrap a piece of masking tape around the cylinder, cutting the tape to a length that is exactly as long as the cylinder's circumference. With a pencil held against the tool rest, manually rotate the turning to mark a line along the length of the tape.

2 **Lay out the facets on the tape.** Carefully remove the masking tape from the turning and lay it out on a flat surface. For a 3-inch-diameter turning, mark a line on the tape every 1½-inches. If your turning is another diameter, divide the length of the tape into six equal parts. A compass set to this radius is a convenient way to make the marks. Each mark represents one of the six corners of the key block.

3 **Transfer the marks to the workpiece, and make the decorative turning.** Rewrap the tape carefully around the turning. Align each of your marks to the tool rest, and scribe a line along the length of the turning. Scribe a sixth line where the ends of the tape meet. Turn the decorative part of the key block.

4 **Plane the facets.** Use a block plane to create flat facets between each pair of lines. You may find it helpful to lay out the depth of the facets along the end of the turning.

Wrap the turning with tape, and cut the tape to the exact circumference.

Use a compass or ruler to divide the tape into six sections of equal length.

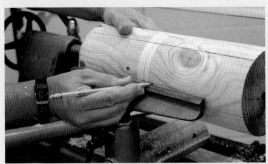

Using the tool rest as a guide, extend the tape marks along the length of the turning.

Plane the six flat facets. Note the depth guide line on the end of the turning.

making a cupola

Here is a basic cupola you can easily assemble yourself in an afternoon. It's made from a few pieces of cedar and a couple of pieces of plywood.

1 Cut the plywood. Cut two pieces of ¾-inch exterior-grade plywood into octagons that measure 21½ inches across opposing corners, as shown in figure 15. On the piece you'll use for the bottom, lay out and cut another octagon 1 inch inside the first.

2 Make and attach the posts. Cut eight pieces of 1½ x 1½-inch cedar, making each piece 10 inches long. Saw meeting 22½° bevels onto the outside face of each post. Attach the posts to the top and bottom pieces with 2½-inch screws.

3 Make and attach the upright. The upright is a 13¾-inch-long piece of 1½ x 1½-inch stock. Center it on the top piece of plywood with a 2½-inch screw.

4 Cut and install the horizontal bands. Rip 1½-inch-thick cedar stock into ⅜-inch slices to make the horizontal bands. Cut them to 8 inches long to fit across the posts. Starting ½ inch up from the bottom of the posts, attach the uprights with 1-inch brads, leaving ½ inch between bands.

5 Cut and install the vertical trim. Cut 16 pieces of cedar to ⅜ x 1½ inches with a 22½° bevel along one edge. Install the vertical trim over the ends of the horizontal bands, attaching them through the bands into the posts with 1½-inch galvanized finish nails.

You can top your gazebo with a homemade cupola. FROM *BUILDING BACKYARD STRUCTURES, SHED, BARNS, BINS, GAZEBOS AND OTHER OUTDOOR CONSTRUCTION* BY PAUL LEVINE, TOM BEGNAL AND DAN THORTON (STERLING PUBLISHING CO. INC, 1997). PHOTO BY PAUL LEVINE.

6 Cut and install the tapered base trim. The base trim is made of ¾ x 2-inch cedar beveled to ⅜ inches, as shown in figure 15. Cut it to fit 8-inch pieces with 22½° angles on the ends to fit along the edge of the lower plywood. Install it under the lowest horizontal band to close the gap between the plywood and the roof.

7 Install the horizontal trim. As shown in figure 15, the horizontal trim overlaps the vertical trim and the base trim. Cut it with 22½° angled ends to fit.

8 Cut and assemble the roof shingles. The roof is covered with eight 10 x 24-inch cedar shingles. Cut the shingles at a length that will reach the top of the center support with an overhang of a couple of inches. After cutting, make marks at the top and bottom of the shingles halfway across their thickness. Then mark the shingle butt 4½ inches to each side of the centerline. Using a metal straightedge, cut the shingle between the top centerline and the marks to both sides of the center of the butt. Assemble the shingles to the cupola with a couple of finish nails into the plywood and a spot of weatherproof adhesive caulk at the center support.

9 **Cut and install the hip caps.** The caps are 3-inch-wide lengths of ½-inch cedar that should end up being about 22 inches long. You'll find it's easier to cut them about 2 feet long, make the angled cuts at top, then mark the bottoms about ½ inch longer than the shingles.

From the middle of one end of each piece, draw 67½° angles to the sides, creating a point. Cut these angles. For all the angled cuts to fit together, you will also need to bevel the pieces on the bottom. Since I don't have a saw that can make that cut, I can't tell you the actual angle, but it is slight. The easiest thing to do is to make the bevels with a block plane, then test fit the pieces until they fit snuggly on the roof. Once you get a fit, mark them for their bottom cuts. If you like, nip back the corners at 45° as shown in figure 15. Attach the hip caps with adhesive caulk. Put a bead of adhesive along the shingle joints and on the shingles where the hip caps will meet. Put a few dabs on the bottom of each cap.

10 **Install the cupola.** Put a bead of adhesive caulk around the bottom of the cupola where it will contact the roof. Drive eight 3-inch screws through the bottom plywood and into the roof.

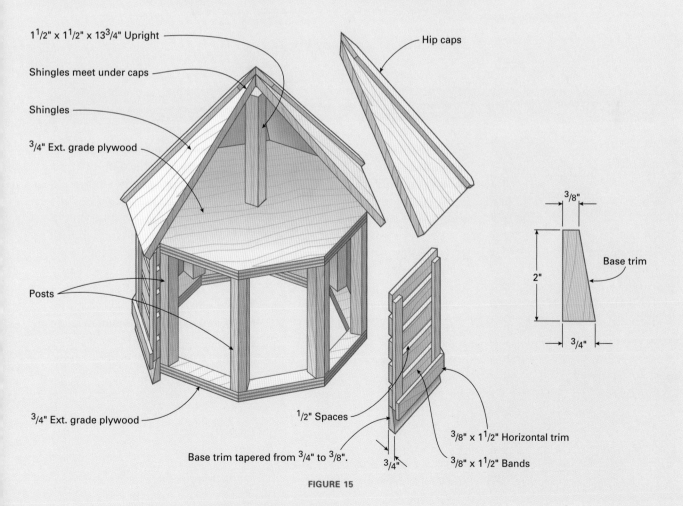

1½" x 1½" x 13¾" Upright

Shingles meet under caps

Shingles

¾" Ext. grade plywood

Hip caps

Posts

¾" Ext. grade plywood

½" Spaces

Base trim tapered from ¾" to ⅜".

¾"

⅜" x 1½" Horizontal trim

⅜" x 1½" Bands

⅜"

2"

¾"

Base trim

FIGURE 15

Rafter Overhangs

Because gazebo rafters are usually exposed under the ceiling, the overhangs are usually left exposed underneath as well. If you like, it's easy to end rafter tails with a flourish by adding a decorative curve or two, as shown in the photo. Just draw a shape that pleases you on the first rafter, along with the bird's-mouth and plumb cuts, and use that rafter as a pattern for the others. Otherwise, you can cut the rafters square or plumb and cover the ends with fascia boards.

Jim Quinlan ended his exposed rafter tails with decorative curves that complement his homemade brackets.

making an enclosed overhang

Both sides of each joist end and rafter tail are beveled in place.

The gazebo we built for this book has ceiling joists that run through the posts to become part of the rafter overhang. For this reason, we enclosed the rafter overhang.

1 Bevel the joist ends and rafter tails. Set your circular saw to 22½° and bevel both sides of each joist end and rafter tail.

2 Install outside 2 x 4s and common rafters. Cut lengths of 2 x 4s and angle the ends to fit between the hip rafters. Toe-screw them into the sides of the hips, flush to three ends. There is one common rafter on each of the gazebo's eight sides. Cut these rafters to fit between the tower window and the outside 2 x 4s. They'll rest on the top plates between the posts, which will place them in the right plane in relation to the hips.

Tongue-and-groove soffit boards are blindnailed in place. The part of the ceiling the soffit will cover is sheathed in plywood, which is cheaper and easier to install than the tongue-and-groove ceiling boards.

3 **Install the fascia.** The fascia boards are 1 x 6 pieces of cedar nailed to the outside 2 x 4s. The fascia have angled ends that meet over the ends of the hip rafters. Using a scrap of wood as a guide, install the fascia slightly lower than the outside 2 x 4, making sure that it isn't higher than the bottom of the roof sheathing.

4 **Install the soffit.** Build the soffit after the roof is in place. In this case, the soffit is made of short lengths of the same tongue-and-groove boards used to sheathe the roof—we stained them blue to match the ceiling. Cut the first piece into a triangle to fit centered under a hip. Nail the soffit boards to the outside 2 x 4 and the top plate.

A wide cedar board closes the inside of the overhang. An L-molding is nailed in place to cover the ends of the soffit pieces.

5 **Close in the overhang.** We closed the opening between the ceiling and the top plate with a piece of ¾-inch-thick cedar. To provide intermediate nailing for these pieces, nail lengths of 2 x 4 to both sides of the common rafters and to the top plate.

If you want, you can carefully cut each tongue-and-groove soffit piece to fit flush to the outside of the top plates. This allows the wide cedar boards to cover the ends of the soffit boards. Daniel, however, chose to let the tongue-and-groove pieces extend ¾ inch beyond the top plate to form a lip for the cedar board to sit on. Then he made L-shaped molding from 1 x 1-inch rips of cedar. In either case, you will need to notch the cedar boards to fit around the common rafters.

building a kicked overhang

Steep roofs give gazebos a dramatic, gothic look. But when slopes get steeper than 12 on 12, you can't get much horizontal overhang without extending the rafters halfway down to the railing. *Kicked rafters* provide an elegant solution that is very easy to construct: You end the full rafters at the top plate. Then, to create the overhang, you attach short kicked rafters at a more gradual slope to each side of the full rafters. Just like any rafter overhang, you can cut kicked rafters square or plumb on the end. You can give them a fascia and/or a soffit or can embellish them with a decorative cut on the ends. Naturally, construction details will vary depending on your design. Shown here is a typical approach that you can easily adapt.

1 Install a continuous band. The continuous band closes up the top of the wall, helps resist the outward thrust of the full rafters, and provides a point of support for the kicked rafters. As shown in figure 16, you nail it across the face of a doubled 2 x 4 top plate. Since the width of the continuous band, the kick-rafter slope, and the overhang length form a right triangle, you can adjust any of these to suit your design. The continuous band shown here was made by ripping 2 x 8s to 6 inches wide. With a 9 on 12 kick-rafter slope, this results in a 16-inch overhang. Bevel the ends of the continuous bands to meet at the hip corners.

2 Cut and install the full rafters. Lay out and cut these rails exactly like rafters with tails. The only difference is that instead of stopping the bird's-mouth plumb cut where it meets the seat cut, you continue the plumb cut to cut the rafter off at that point. Common rafters will butt squarely into the continuous band. Hips will meet the bands at the corners, leaving a small gap—about ⅜ inch for 1½-inch-thick rafters. To allow for this gap, cut the hips ⅜ inch shorter.

A roof with kicked rafters is a graceful way to create an overhang on a steep roof.
PHOTO BY DAVID SCHIFF.

3 **Make and attach the kick rafters.** Using your framing square, draw a line across a piece of rafter stock at the slope you have chosen for the kick rafters. Put the piece of stock against the side of an installed full rafter and slide it down the continuous band until you have the overhang you want. Align a level to the slope line you drew, and adjust the rafter until the line is plumb. Check the overhang length and adjust it if you like. When you are satisfied, scribe along the full rafter to mark the top cut on the kick rafter. Make the cut and use this for your kick rafter pattern. Just as hip slopes are different than common slopes, hip kicks differ from common kicks. Because of this, you will have to use this procedure to make separate patterns for both. Attach the kicks to the rafters with four 2½-inch screws.

4 **Add optional fascia, soffit, and decorative molding.** The kicks shown in figure 16 are cut square on the ends and covered with a 1 x 8 soffit board. Note that the soffit has an angle cut on one end. Screw this end to the fascia. Attach the other end to the continuous band with galvanized finish nails. You may find that a decorative molding will be a nice touch for your design. Using the molding also has a practical advantage since it moves the roof edge a bit farther out from the fascia.

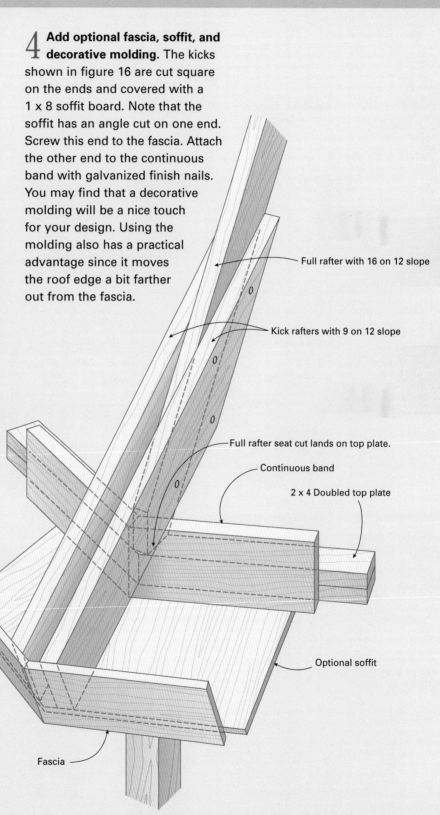

Full rafter with 16 on 12 slope

Kick rafters with 9 on 12 slope

Full rafter seat cut lands on top plate.

Continuous band

2 x 4 Doubled top plate

Optional soffit

Fascia

FIGURE 16

making curved rafters

Don't you just love tricks of the trade that make a seemingly complex detail really simple? Of course you do, and that's why you'll love this method for making curved rafters. Forget about trying to lay out a curve by connecting points on a grid. Instead, use a simple bending batten, a technique boat builders use to lay out curved sides.

There's just one structural concern when you use curved rafters: The thickness of the rafter at the deepest point is the effective structural thickness of the rafter. For example, in the photo below, a 3½-inch-deep curve is laid out on a 2 x 8 rafter, leaving a thickness of 3¾ inches at the center. This means that structurally this rafter can only carry the same load as a 2 x 4.

Curved rafters are easy to make with a technique borrowed from boat builders.
PHOTO BY DICK DIETRICH.

1 Make the batten. Start by ripping a ½-inch-thick batten from clear, straight-grained pine or cedar. The grain is important, because knots or wild grade will distort the curve. Make the batten a foot or so longer than the rafter. The width of the batten is unimportant.

2 Lay out the center point. Start by laying out the top and bottom plumb cuts of the rafter. Then draw a centerline across the width of the rafter. The line should be centered along the length of the rafter as measured from the top of the top plumb cut and the top of the bottom plumb. Along the centerline, measure in from the crown to the maximum depth of the curve. Drive a finish nail halfway into the board at this point. Also drive nails into the plumb lines approximately ½-inch in from the crown edge of the rafter.

3 Bend the batten and draw the curve. Place the middle of the batten behind the center nail, and pull the ends in front of the end nails. If necessary, you can bend the end nails a little so the concave edge of the batten exactly crosses the ends of the plumb cuts. Draw the curved cut line along the waste side of the batten. Remove the batten and nails, and you're ready to make the cut with a saber saw.

The batten is bent over the center nail and behind the end nails.

Sheathing and Roofing

Probably the most common way to cover a gazebo roof is to sheathe it with tongue-and-groove boards that will remain exposed as the ceiling, then cover them with cedar shingles or shakes for the roof.

The other do-it-yourself roofing option is asphalt shingles. While it is possible to nail asphalt shingles directly to tongue-and-groove boards, this is not advisable, and the shingle warranty may be voided unless you nail the shingles to plywood sheathing. The reason for this is that solid wood can move, shrink, and expand enough from changes in humidity, to cause asphalt shingles to split over time. Therefore, if you want a tongue-and-groove ceiling with an asphalt roof, you'll need to cover the boards with a layer of ½-inch CDX grade plywood before roofing. (See pages 38 and 39 for more on roofing materials.)

If you are planning to use curved rafters, (see Making Curved Rafters on page 143), tongue-and-groove sheathing is your best choice. You can angle the joint between each board enough to make the gradual curve that is typical of most gazebo roofs. Half-inch plywood will bend enough to make very gradual curves. You can also use cut ¼-inch-deep kerfs across the tops of the plywood sheets to make them conform to the curve.

Applying asphalt shingles to a gazebo roof is pretty much the same as for any hip roof. Since the basic procedures are well covered in inexpensive paperback how-to books that you find at any home center or lumberyard, I won't waste precious space leading you through them here.

However, the one problem most roofing books won't address is how to finish the peak of six or eight converging hips. As far as I can tell, there is no really good way to do this with asphalt. With cedar shingles or shakes you could carefully craft a six- or eight-sided cap, gluing the edges together with a waterproof exterior glue such as polyurethane. But after all the sweat and creativity you poured into this gazebo, do yourself a favor: Either top the gazebo with a cupola or visit your local sheet metal shop and have them fabricate a simple, handsome copper cap.

 ## roofing with cedar shingles or shakes

The procedures for installing shingles and shakes on a gazebo are essentially the same as for any hip roof and are readily available in roofing how-to books. The work is a little trickier though, so I will take you through the installation of shakes on the gazebo that we built for this book.

1 **Install the sheathing.** As mentioned earlier, since the gazebo featured at the back of this book has a closed overhang, we covered the portion of the roof that won't be seen with plywood sheathing to save time and money. Then we contiued the ceiling with tongue-and-groove boards, then covered the entire ceiling with a layer of ¾-inch plywood. This provided a 1½-inch-thick surface so the shakes could be nailed securely without having the nails poke through the ceiling.

By now you are intimately aware of the angles involved in constructing your gazebo to realize that you would have to bevel the edges of sheathing boards and/or sheets of plywood to have them butt fully together at the hips. Don't bother; just cut the edges square. It's fine for the joints to form open Vs since they will be covered.

The lower part of the ceiling is plywood. Tongue-and-groove boards take over where they will be visible in the ceiling. The entire ceiling is then covered with plywood.

To bend drip the edge smoothly around a corner, first snip the bottom leg at the center of the hip, as shown in the photo above. Then snip the top leg and secure the overlap with a single roofing nail, as shown in the photo below.

2 Install drip edge and felt paper. The edges of the plywood are covered with 10-foot lengths of aluminum drip edge. Wrap the drip edge around the corners, securing it with a roofing nail through the top leg every 12 inches or so. Place the nails 1 to 2 inches from the top edge. One snip on the bottom leg and one on the top allows the drip edge to bed smoothly around the corners. Staple the underlayment to the roof, starting at the bottom and overlapping joints at least 2 inches.

3 Install starter shingles and first shakes. You can use starter shingles for both shingle and shake roofs. Let the starters overhang the drip edge by about ¾ inch and space them about ½ inch apart to allow for humidity movement. For each course, start by installing the shakes that meet at the hips. These shakes meet at an angle. Like the sheathing, these joints don't need to be perfect. Make sure the shakes cover starter shingle joints by at least 1½ inches-this same rule applies for subsequent courses of shakes. Extend the shakes about 1½ inches past the drip edge. Use 6d hot-dipped galvanized aluminum or stainless-steel ring-shank nails for each shake. For this gazebo, we overlapped shake courses to leave an 8-inch exposure. In order for the nails to be covered by the next course, we nailed the shakes approximately 9 inches from the bottom.

Start each course with shakes cut at an angle to meet at the hip.

4 Install subsequent courses. Continue installing shakes up the gazebo roof, always starting at the hips and fitting shakes between. A utility knife is all you need to split shakes down their length. I like to vary the exposure by ½ inch or so from shake to shake. I think that complements the rustic look of shakes better than perfectly aligned butts. You may want to make a story pole marked with the exposure for several courses to guide your installation.

A story pole helps gauge how much of each shake is exposed to the weather.

5 **Install hip caps.** Premade hip-cap units are available, but they won't fit the angles on six- or eight-sided gazebos. The most common solution is to assemble your own caps on site, as shown in figure 17. Use a sliding T-bevel laid over the hip to find the bevel angle for the edges of the shingles or shakes, then bevel cap shingles or shakes on the table saw.

Begin with a starter course that is as long as the weather exposure, and be sure to overlap the joints. Use nails that are long enough to go through the shingles or shakes below and at least ½ inch into the sheathing.

For the gazebo we built for this book, Daniel Grant decided to make his caps of rough-sawn cedar boards. The boards have a different texture than the split shakes, creating a nice design detail.

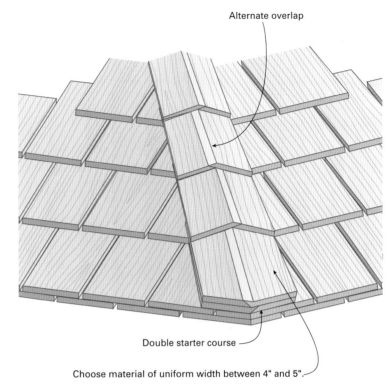

Alternate overlap

Double starter course

Choose material of uniform width between 4" and 5".

FIGURE 17. Hip and Ridge Application

do you need a lightning rod?

Daniel Grant's gazebo that he built for this book is literally out standing in its field. There are no other structures or trees nearby, so the tall gazebo is a prime target for lightning. To protect against lightning damage, Daniel mounted a lightning rod into the key block. To act as a ground for deflecting a strike, Daniel ran a copper cable from the lightning rod to another metal stake in the ground.

A lightning rod extends from the key block with copper cable running to a metal stake in the ground. PHOTO BY DANIEL GRANT.

CHAPTER NINE

9

One Gazebo Start to Finish

Tʜɪs chapter has two purposes. First, it provides you with the specific plans and dimensions you'll need if you would like to construct the gazebo Daniel Grant built for this book. Second, it gives you a concise order of procedure—a snapshot of steps—for building your own gazebo design. Of course you may decide to build a hybrid—perhaps you like the overall dimensions but have your own idea for railings.

As an example, if you like the style of this gazebo, but aren't looking for a 20-foot-tall structure with a tower, it would be very easy to adapt this design to eliminate the tower. Just continue the bottom rafters to a key block or perhaps add a cupola. (See Making Key Blocks, page 132, or Joining Rafters to Headers, page 123, and Making a Cupola, page 136.) Since you won't need a tower platform, bring the ceiling joists together in the middle using a half-lap joint where two full-length joists cross and bring four half-length joists to the center with 45° angles. Without the tower, you might like the roof better if you increase its slope a bit.

I have also compiled a parts list in case you want to build this exact gazebo. Of course you'll use the part lengths given in the list as a guide only—in any carpentry project it always makes sense to measure and cut to fit as you go. If you are building your own design, use the list as an example for a checklist to make sure you have ordered all the materials you need.

What you won't find in this chapter is information about building techniques. Instead you will find page references back to the pertinent techniques discussed in chapters four through eight.

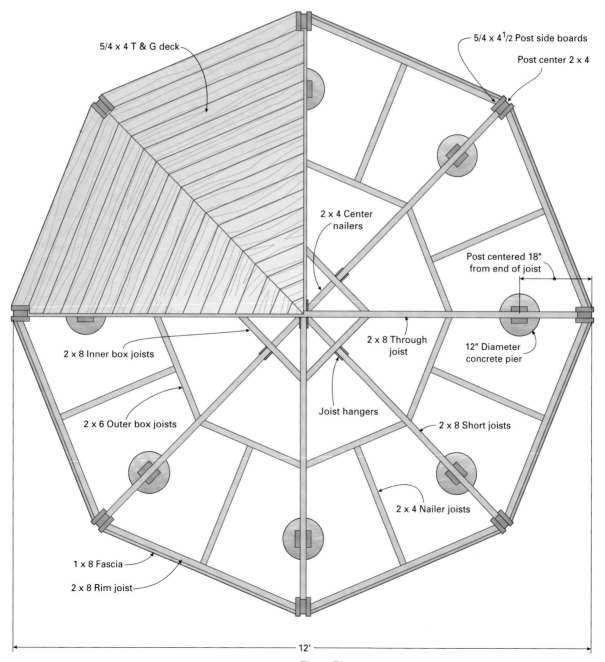

FIGURE 1. Floor Plan

5/4 x 4 T & G deck

5/4 x 4 1/2 Post side boards

Post center 2 x 4

2 x 4 Center nailers

Post centered 18" from end of joist

2 x 8 Inner box joists

2 x 8 Through joist

12" Diameter concrete pier

2 x 6 Outer box joists

Joist hangers

2 x 8 Short joists

2 x 4 Nailer joists

1 x 8 Fascia

2 x 8 Rim joist

12'

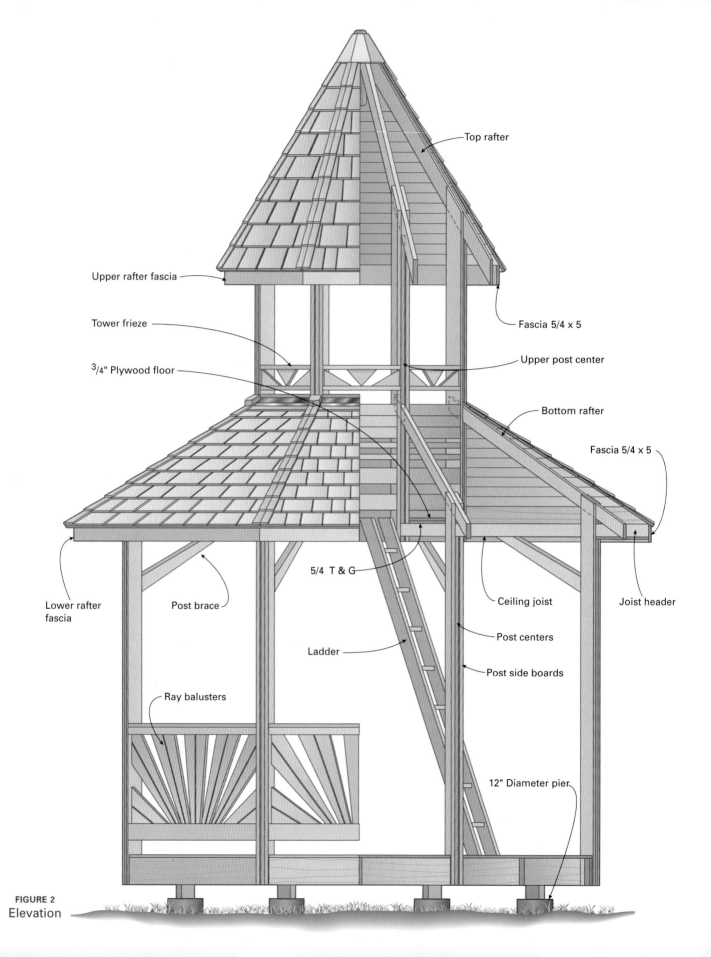

Top rafter

Upper rafter fascia

Fascia 5/4 x 5

Tower frieze

Upper post center

³/4" Plywood floor

Bottom rafter

Fascia 5/4 x 5

5/4 T & G

Lower rafter fascia

Post brace

Ceiling joist

Joist header

Post centers

Ladder

Post side boards

Ray balusters

12" Diameter pier

FIGURE 2
Elevation

1 **Lay out the piers.** Lay out the center point of the eight piers. Note that the joists will be cantilevered so the center of each pier will be 18 inches in from the joist ends. (See Laying Out an Eight-Sided Gazebo, page 47.)

2 **Dig and pour the footings.** Dig footings to the appropriate frost-depth for your area. Insert round forms and fill with concrete. (See Poured Concrete Footings, page 49.) Insert J-bolts into the center of the piers before the concrete sets.

3 **Frame the floor.** (See Concentric Joist Framing, page 62, and Cantilevering the Concentric Joist Plan, page 65.) Don't install the rim joists until after the posts are in place.

4 **Construct and attach the posts.** The posts are built up from two 5/4 x 4½-inch pieces with a 2 x 4 sandwiched between. Assemble them and attached them to the joists as described in Cantilevering the Concentric Joist Plan, steps 7 and 8. Install the outer rim joists and the fascia.

5 **Install the floor.** Install the 5/4 x 4 tongue-and-groove porch flooring concentrically as described in Installing Concentric Flooring, page 70.

6 **Install the center insert.** Daniel Grant used a ceramic turtle for his center insert; you'll probably choose something different. (See Making a Center Medallion, page 72.)

7 **Build steps or platform, if necessary.** The need for steps or a platform depends on your site, so no platform or step is included in the plan here. (See Building Stairs, page 74 and Building an Entrance Platform, page 80.)

8 **Install the ceiling joists and build the tower floor.** The ceilings joists consist of 2 x 4s that run through each post and meet at the middle to support a platform for the tower. After installing the joists, cut them off 11 inches from the post side boards to create an overhang. Install horizontal 2 x 4s between the posts and in the same plane as the joists. (See Installing Ceiling Joists, page 86, Ceiling Joists Support a Tower Floor, page 89, and A Variation on Sideboard Construction, page 97.) Install the tongue-and-groove ceiling under the platform leaving one section open as shown in the photo below. Install the seven blocks around the perimeter of the floor.

When you install the upper rafters, you will be looking up most of the time while working on the platform. This makes it easy to forget, and you could step into the hole. To prevent this, cut the plywood floor to size without cutting away the opening section. Tack it in place with a few screws. After the rafters are installed, scribe the opening on the plywood, remove it to make the cuts and then install it permanently.

One section of the tower floor is left open to allow access for the ladder. Note that two of the joists are notched to accept the ladder rails. PHOTO BY DAVID SCHIFF.

9 **Cut and install the post braces.** The post braces help prevent lateral movement of the tower and also add an attractive detail to the inside of the gazebo. If you build the gazebo without the tower, you can skip the braces if you like. Cut the braces from a 22-inch length of cedar 2 x 4, tapering it as shown in figure 3 and angling the ends at 30° and 60° as shown. Install them wide side down with a 6-inch lag screw into the posts and a 3½-inch lag screw into the ceiling joist. Countersink the screws and cover with a wooden plug.

10 **Make and install the railings.** Make patterns and cut and install the railings as described in detail in Making a Sunburst Railing on page 110.

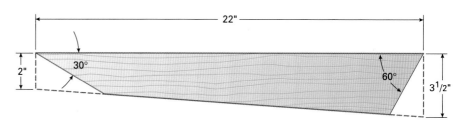

FIGURE 3
Making the Post Brace

11 **Install tower posts, hip rafters, and common headers.** Cut center 2 x 4s for the tower posts to length with 60° angles on one end. Cut the lower hip rafters to length with 30° angles. Assemble the rafters to the posts as described in Sketching in the Sky on page 130. The headers also serve as the bottom of the window. Each header consists of three 2 x 4s installed with their faces vertical. Install the headers between the tower posts, aligning the bottom of the headers to the height that the bottom rafters meet the posts. Note in the photo below that the bottommost 2 x 4 over the opening is tapered to increase headroom as you climb the ladder to the tower.

stealing a little headroom

After climbing up the tower ladder for the first time, Daniel Grant decided just a little more headroom was needed. So he got out a grinder and went to work on the bottom 2 x 4 header over the opening. Cedar is soft so it didn't take long to taper back the 2 x 4 and add a little upward curve. The end of the common rafter got ground down a bit to meet the header's new shape. Such spur-of-the-moment solutions have the bonus of adding to the gazebo's organic, handcrafted appeal.

The header and rafter over the ladder opening got some custom treatment to add a bit of needed headroom. PHOTO BY DAVID SCHIFF.

12 **Install the top rafters and top post 2 x 4s.** Make a key block as described in Octagonal Key Blocks on page 132. Cut the top rafters with 60° degree plumb cuts at top and bottom and install as described in Assembling the Rafters to the Key Block on page 135. Cut the top post center 2 x 4s to length with 30° angles on one end and install them under the top rafters.

13 **Assemble the rafter over-hangs and common rafters.** Install the eight outside 2 x 4s running between the rafter tails. Then install the common rafters between these outside 2 x 4s and the common headers. Add the fascia boards. Close in the soffit and inside of the overhang as described in Making an Enclosed Overhang on page 138. The tower overhangs only have fascia boards.

14 **Sheathe the roof and cover with shakes.** See Sheathing and Roofing, page 142. Cover the top of the last course of shakes with 1 x 8 boards that surround the tower as shown in the photo at right. Bend 4-inch copper flashing at the middle to lap the boards and the side of the tower. Lap a full ridge piece over the 1 x 8 and the flashing, then add a half-length ridge piece on top. Finally, add 2 x 4 blocks that are angled at front and notched to fit into the tower posts. The blocks are routed out to fit over the flashing.

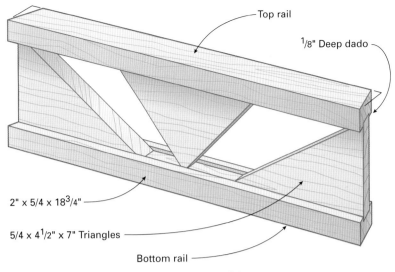

Top rail

1/8" Deep dado

2" x 5/4 x 18³/₄"

5/4 x 4¹/₂" x 7" Triangles

Bottom rail

FIGURE 4. The Tower Frieze

15 **Make and Install the frieze.** The tower frieze rail has a three-piece sunburst that consists of three triangles made of 5/4 x 4½-inch stock dadoed into top and bottom rails made of 5/4 x 2-inch pieces as shown in the figure 4. The ⅛-inch deep dado can be made with a router or with several passes over a table saw. Angle the ends of the friezes to meet the tower posts.

A ring of 1 x 8s, copper flashing, and notched blocks complete the roof details. PHOTO BY DAVID SCHIFF.

16 **Make the ladder.** For light weight and strength, the ladder to the tower is made of the 5/4 mahogany flooring stock ripped down to 3 inches wide and rounded over on the ends. There are eight rungs dadoed into the 10-foot-long rails. Seven of these rungs are spaced 12 inches apart, starting 12 inches from the floor. The last rung is dadoed in place 1 inch from the top—the top 3 feet of the ladder serves as a handrail as you step onto the platform. The rails and dadoes are cut at an angle of about 75° to give the ladder some slope.

tower jump seats

The gazebo tower features seven jump seats that fold up and down and are removable. This adds a lot of versatility to use of the tower space. The seats are as simple to make as they are ingenious in design.

1 Make the seat boards. Each jump seat has two 5/4 x 4½-inch cedar seat boards. Cut them to 16 inches long. Cut 67 ½° (22½° from 90° degrees) angles on each front seat board.

2 Cut and install the cleats and crosspieces. All of these pieces are made from 1⅜-inch rips of 5/4 stock. For each seat, cut two cleats to 14 inches and two crosspieces to 10½ inches. Screw the top crosspiece across the ends of the cleats and the bottom cross piece 1 inch in from the ends as shown in figure 4. Use 2½-inch screws. Round the edges of all pieces with a router.

3 Install the seat boards. Center the angled front seat board across the cleats and flush to the front of the cleats, then screws the cleats to the board. Using a 2 x 4 spacer, position the back seat board 1½ inches from the front board and screw it in place.

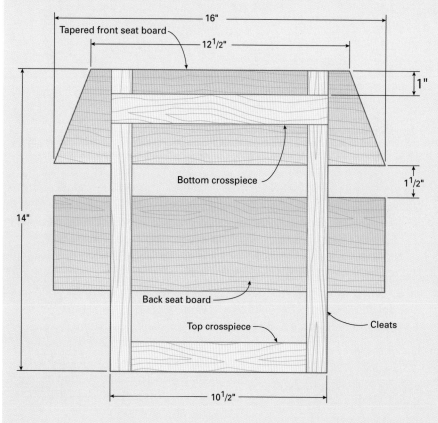

- 16"
- Tapered front seat board
- 12½"
- 1"
- Bottom crosspiece
- 1½"
- 14"
- Back seat board
- Top crosspiece
- Cleats
- 10½"

FIGURE 4

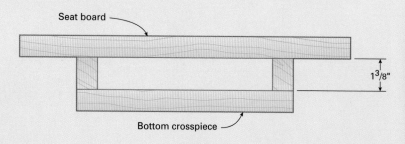

- Seat board
- 1⅜"
- Bottom crosspiece

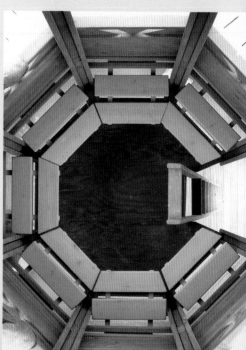

In the open position the top crosspieces hook behind the lower narrow rail while the cleats rest on the upper wide rail. PHOTO BY DANIEL GRANT

In the upright position, the lower narrow rail fits between the seat boards and the cleats rest against the upper wide rail.
PHOTO BY DAVID SCHIFF.

4 Install the jump seat rails. The jump seat rails close in the bottom of the tower while providing storage and support for the seats. For each seat, two of the rails are 4½ inches wide and two are 1½ inches wide. Figure 5 shows their lengths and positions between the tower posts. Attach them with angled 3-inch screws.

5 Install the seats. To install, slip the jumpseats between the higher of the two 4½-inch-wide rails and the 1½-inch-wide rail just above it. In storage mode the front seat board fits between the two narrow rails and the cleats rest against the upper wide rail as shown in the above photo. In the open position the top crosspiece hooks behind the lower narrower rail while the cleats rest on the upper wide rail.

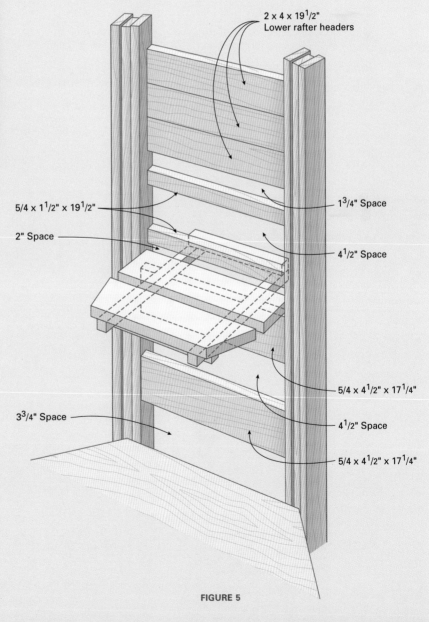

2 x 4 x 19½"
Lower rafter headers

5/4 x 1½" x 19½"

2" Space

1¾" Space

4½" Space

5/4 x 4½" x 17¼"

3¾" Space

4½" Space

5/4 x 4½" x 17¼"

FIGURE 5

parts list

QUANTITY	PART	DIMENSIONS	MATERIAL
floor frame			
8	posts	4 x 4 x as needed	pressure-treated
1	through joist	2 x 8 x 12 ft.	pressure-treated
2	half-through joists	2 x 8 x 11 ft. 10½ in.	pressure-treated
8	inner box joints	2 x 8 x 22 in.	pressure-treated
4	short joists	2 x 8 x 10 ft.	pressure-treated
8	outer box joists	2 x 6 x 22½ in.	pressure-treated
8	rim joists	2 x 8 x 52 in.	pressure-treated
8	nailer joists	2 x 4 x 66 in.	pressure-treated
8	center nailers	2 x 4 x 18¾ in.	pressure-treated
8	fascia	2 x 8 x 52½ in.	cedar
400 linear ft. of 5/4 x 4 in. tongue-and-groove flooring			
bottom posts			
8	post centers	2 x 4 x 8 ft.	cedar
16	post side boards	5/4 x 4½ x 9 ft.½ in.	cedar
railings			
300 linear ft. ray balusters		5/4 x 4½ in.	cedar
16	inner top and bottom bands	5/4 x 2 x 50½ in.	cedar
16	outer top and bottom bands	5/4 x 2 x 52½ in.	cedar
8	cap rails	2 x 4 x 53 in.	cedar
8	subrails	5/4 x ⅞ x 53 in.	cedar
ceiling joists and platform			
1	through joist	2 x 4 x 14 ft. 1½ in.	cedar
2	half-through joists	2 x 4 x 84 in.	cedar
4	inner box joists	2 x 4 x 22 in.	cedar
4	short joists	2 x 4 x 72¾ in.	cedar
1	platform floor	¾ x 4 sq. ft.	pressure-treated plywood
7	platform perimeter supports	2 x 4 x 16¼ in.	cedar
7	floor edge trim	5/4 x 1½ x 17 in.	cedar
tower posts and rafters			
16	upper post side boards	5/4 x 4½ x 8 ft. 2 in.	cedar
8	lower post centers	2 x 4 x 31 in.	cedar
8	upper post centers	2 x 4 x 52½ in.	cedar
8	lower hip rafters	2 x 4 x 61½ in.	cedar
24	lower rafter headers	2 x 4 x 19½ in.	cedar
8	lower post headers	2 x 4 x 52 in.	pine or fir

QUANTITY	PART	DIMENSIONS	MATERIAL
tower posts and rafters, *continued*			
8	outer soffit supports	2 x 4 x 64 in.	pine or fir
8	common rafters	2 x 4 x 57½ in.	cedar
1	key block	4 x 4 x 20 in.	pressure-treated
8	upper rafters	2 x 4 x 70 in.	cedar
8	upper rafter fascia	5/4 x 4½ x 29½ in.	cedar
8	lower rafter fascia	5/4 x 4½ x 65¾ in.	cedar
16	soffit nailing blocks	2 x 4 x 10½ in.	pine or fir
8	inside overhang closure	1 x 12 x 49½ in.	cedar
8	L-molding for inside closure	1 x 1 x 49½ in.	cedar
14	jump-seat crosspieces	5/4 x 4½ x 17¼ in.	mahogany
14	jump-seat cross pieces	5/4 x 1¼ x 19½ in.	mahogany

tower frieze

QUANTITY	PART	DIMENSIONS	MATERIAL
16	frieze rails	2 x 4 x 19½ in.	cedar
24	frieze triangles from	5/4 x 4½ in.	cedar

ladder

QUANTITY	PART	DIMENSIONS	MATERIAL
2	rails	5/4 x 3 x 10 ft.	mahogany
8	rungs	5/4 x 3 x 11 in.	mahogany

jump seats

QUANTITY	PART	DIMENSIONS	MATERIAL
14	seat boards	5/4 x 4⅜ x 16 in.	cedar
14	cleats	5/4 x 1⅜ x 14 in.	cedar
14	crosspieces	5/4 x 1⅜ x 10½ in.	cedar

ceilings and soffits

750 lineal feet of 5/4 x 4 tongue-and-groove		pine

roofing materials

Three 4 x 8 sheets of ¾-in. CDX plywood for sheathing

One roll 15# felt underlayment

17 bundles of 18-in. heavy hand-split shakes

200 lineal ft. of 1 x 8 cedar boards for ridge caps and tower flashing

20 ft. of 4-in.-wide copper flashing

metal connectors

Eight 4 x 4 post bases

Four 2 x 8 joist hangers

Acknowledgments

It takes a team of people to create any book, especially a book like this one that requires careful coordination of text, photographs, and illustration. First, I'd like to thank my closest compatriots in this work, Daniel Grant and his wife Ingela Noren. Daniel designed and built the featured gazebo and contributed many of the design ideas and construction techniques in the book. Ingela cheerfully and graciously tolerated the constant interruptions to her life as an artist and mother to take the hundreds of photos that document the building process. Daniel and Ingela produce unique art furniture and finials. You can check them out at Grant-Noren.com.

It was a lucky day when I was introduced to Jim Quinlan just as he was building a gazebo in his backyard. Jim, a shop teacher at Vernon Township High School in Vernon, New Jersey, gave me hours of his time and invited me into the school workshop where he demonstrated techniques for Ingela's camera. Jim did this for nothing more than the love of sharing knowledge. The fact that his approach to building is very different than Daniel's added greatly to the depth of this book.

Thanks to May Brawley Hill, whose book *Furnishing the Old-Fashioned Garden, Three Centuries of Summerhouses, Dovecotes, Pergolas, Privies, Fences & Birdhouses* (Harry N. Abrams, Inc., 1998) was the single most useful source in giving me a perspective on the history of gazebos. Thanks also, May, for graciously inviting my wife Beth and me, total strangers, to your home for a lovely lunch and interview.

Rudy J. Favretti, author of *Landscapes and Gardens for Historic Homes* (AltaMira Press, 1997) also was kind enough to share his knowledge with me. Thank you Prof. Favretti.

When I was scratching my head over the math involved in gazebo roofs and how best to explain it, I knew the place to turn was my old friend and master carpenter Douglas Goodale. Less than an hour at his dinner table helped me clarify the concepts.

Thanks also to landscape architect David Delardi of Landscape Perceptions in Wayne, New Jersey, and Patricia Lundquist of Bellingham, Washington, who shared drawings and ideas about how they sited gazebos into the landscape. I would also like to thank Tim Bullock of Bullock & Company, Ontario, Canada, for sharing his log gazebo design and Marvin Davis of Romancing the Woods in Woodstock, New York, who contributed photos and historic information about rustic gazebos.

At Lark Books I would like to thank my old friend and new editor Andy Rae, who was a sounding board when I was getting mired. When Andy took off to write his own darn books, editor Jane LaFerla hit the ground running and did a great job of tying up loose ends. Thanks too to assistant editors Veronika Alice Gunter and Anne Hollyfield for a terrific job of photo research.

My thanks to another old friend, illustrator Frank Rohrbach, for turning my rough sketches into beautiful, informative drawings. Frank, you've only gotten better since we last worked together 15 years ago.

And finally to my wonderful wife Beth Kalet and my daughters Aliza and Hannah Schiff, thanks for putting up with my bellyaching through the long arduous process of making this book.

David Schiff
Warwick, New York

Metrics Conversion Chart

length

US	METRIC	US	METRIC
⅛	3 mm	12	30
¼	6 mm	13	32.5
⅜	9 mm	14	35
½	1.3	15	37.5
⅝	1.6	16	40
¾	1.9	17	42.5
⅞	2.2	18	45
1	2.5	19	47.5
1¼	3.1	20	50
1½	3.8	21	52.5
1¾	4.4	22	55
2	5	23	57.5
2½	6.25	24	60
3	7.5	25	62.5
3½	8.8	26	65
4	10	27	67.5
4½	11.3	28	70
5	12.5	29	72.5
5½	13.8	30	75
6	15	31	77.5
7	17.5	32	80
8	20	33	82.5
9	22.5	34	85
10	25	35	87.5
11	27.5	36	90

length

U.S.	MULTIPLY BY	METRIC EQUIVALENT
Foot	0.3048	Meter
Yard	0.9144	Meter
Mile	1.609344	Kilometer

area

U.S.	MULTIPLY BY	METRIC EQUIVALENT
Square inch	645.16	Square millimeter
Square foot	0.09290304	Square meter
Square yard	0.8361274	Square meter
Square mile	2.589988	Square kilometer
Acre	0.40469	Hectare

mass

U.S.	MULTIPLY BY	METRIC EQUIVALENT
Ounce	0.02834952	Kilogram
Pound	0.45359237	Kilogram
Ton	0.9071847	Tonne

volume

U.S.	MULTIPLY BY	METRIC EQUIVALENT
Fluid ounce	29.57353	Milliliter
Gallon	3.785412	Liter
Cubic inch	16.387064	Cubic millimeter
Cubic foot	0.02831685	Cubic meter
Cubic yard	0.7645549	Cubic meter

temperature

FORMULA

Degrees Fahrenheit
Minus 32
Times 5
Divided by 9
Equals degrees celsius

pressure

PSI	MULTIPLY BY	EQUALS
Pounds per square inch	6.896	Kilopascals

Index